Economics of Water Management in Agriculture

Economics of Water Management in Agriculture

Editor

Pankaj K. Dubey

scitus
academics

Economics of Water Management in Agriculture

Edited by **Pankaj K. Dubey**

ISBN: 978-1-68117-021-3

Library of Congress Control Number: 2015931523

SCITUS Academics LLC,
www.scitusacademics.com
Box No. 4766, 616 Corporate Way,
Suite 2, Valley Cottage,
NY 10989

Notice

Reasonable efforts have been made to publish reliable data and views articulated in the chapters are those of the individual contributors, and not necessarily those of the editors or publishers. Editors or publishers are not responsible for the accuracy of the information in the published chapters or consequences of their use. The publisher believes no responsibility for any damage or grievance to the persons or property arising out of the use of any materials, instructions, methods or thoughts in the book. The editors and the publisher have attempted to trace the copyright holders of all material reproduced in this publication and apologize to copyright holders if permission has not been obtained. If any copyright holder has not been acknowledged, please write to us so we may rectify.

Printed in United States of America on Acid Free Paper ∞

Contents

Preface

This book outlines the current trends in the economics of water management in agriculture. It also addresses some of the most relevant current and perspective issues for the water policy offering a wide variety of innovative approaches and original and relevant cases.

Water management is the management of water resources under set policies and regulations. Water, once an abundant natural resource, is becoming a more valuable commodity due to droughts and overuse. Irrigation can improve crop production, reduce yield variability, and increase profits. But choosing and buying an irrigation system are both expensive and complex. When considering investing in an irrigation system, farmers must keep in mind several major factors: financing, crop mix, energy prices, energy sources, commodity prices, labor availability, economies of scale, the availability of water, savings in field operations, the system's application efficiency, the operating pressure of the design, and the depth from which the water must be pumped, or pumping lift.

It reports recent economic research on the theory, practice and policies of water management and investigates problems on the economics of water management in an agricultural context. Specific case studies, methods and techniques in water management are also considered.

Editor

Chapter 1

Improving Water-Efficient Irrigation: Prospects and Difficulties of Innovative Practices

Les Levidow [a], Daniele Zaccaria [b], Rodrigo Maia[c], Eduardo Vivas [c], Mladen Todorovic [d], Alessandra Scardigno [d]

[a] Development Policy and Practice, Open University, UK, Milton Keynes MK7 6AA, UK

[b] University of California, Davis, CA, United States

[c] Universidade do Porto, Prac¸ a de Gomes Teixeira, Portugal

[d] CIHEAM—Mediterranean Agronomic Institute of Bari (IAMB), Bari, Italy

ABSTRACT

Innovative irrigation practices can enhance water efficiency, gaining an economic advantage while also reducing environmental burdens. In some cases the necessary knowledge has been provided by extension services, helping farmers to adapt and implement viable solutions, thus gaining more benefits from irrigation technology. Often investment in technological improvements has incurred higher water prices, however, without gaining the full potential benefits through water efficiency. Farmers generally lack adequate means and incentives to know crops' water use, actual irrigation applications, crops' yield response to different water management practices, and thus current on-farm water-efficiency levels.

Those general difficulties are illustrated by our two case studies investigating options, stimuli and difficulties to improve water-efficient practices. The two areas have strong stimuli for improvement but lack a knowledge-exchange system to help farmers and resource managers identify scope for improvements. Partly for this reason, farmers' responsibility for efficient water management has been displaced to hypothetical prospects, e.g. extra supplies from reuse of treated wastewater or a long-term low water pricing. In both cases a displaced responsibility complements the default assumption that farmers' irrigation practices already have adequate water-use efficiency. Under current circumstances, agricultural water management will maintain the unknown water-efficiency level and farmers will have weaker incentives to make efforts for more efficient practices. A continuous knowledge-exchange is necessary so that all relevant stakeholders can share greater responsibility across the entire water-supply chain. On this basis, more water-efficient management could combine wider environmental benefits with economic advantage for farmers.

INTRODUCTION

Irrigation systems have been under pressure to produce more with lower supplies of water. Various innovative practices can gain an economic advantage while also reducing environmental burdens such as water abstraction, energy use, pollutants, etc. (Faurès and Svendsen, 2007).

Farmers can better use technological systems already installed, adopt extra technologies, enhance their skills in soil and water management, tailor cropping patterns to lower water demand and usage, reduce agrochemical inputs, etc. Water-efficient practices potentially enhance the economic viability and environmental sustainability of irrigated agriculture, without necessarily reducing water usage. To inform such practices, experts have developed various models of water efficiency, yet these are little used by farmers.

Through two case studies in the EU context, this paper will address the following questions:

- When an irrigation area invests in innovative technology, how can its operation help farmers to achieve the full potential benefits together, e.g. an economic advantage, greater water-use efficiency and lower resource burdens?
- Why are innovative technologies often applied in ways which miss the full potential benefits?
- What tensions arise among various objectives and potential benefits?
- How can these difficulties be addressed?

The paper first surveys analytical perspectives on irrigation efficiency – especially the means, incentives and limitations – as a basis to analyse two cases and draw general conclusions.

INNOVATIVE IRRIGATION PRACTICES: ANALYTICAL PERSPECTIVES

Innovative irrigation technology is generally promoted as raising water-use efficiency along with multiple benefits, but these remain elusive in practice, as outlined in the first sub-section below. The limitations have fundamental reasons, as outlined in Section 2.2. To address these issues, our case studies are introduced in Section 2.3.

Practical Limitations of Water-Efficient Irrigation Technology

EU policy frameworks place great expectations upon technologies to improve water efficiency. The European Commission emphasises 'technological innovation in the field of water, given that water efficiency will be an increasingly important factor for competitiveness' (CEC, 2008). According to the European Parliament, solutions should be found in 'clean technologies that facilitate the efficient use of water' (EP, 2008).

Such technological expectations arise in expert reports on agricultural water use:

"Water-efficient irrigation, irrigation on demand and irrigation using brackish water are technologies that will enable the better husbandry of more scarce freshwater resources. Technological developments in respect of irrigation will encompass sensors and communication, intelligent watering systems and high-efficiency delivery mechanisms for water and nutrients, as well as the means of incorporating all of these elements into irrigation 'packages' (EIO, 2011: 25)."

Likewise water efficiency can be enhanced by better using current installations and/or by adopting new equipment (WssTP, 2012: 9).

The main European farmers' organisation has likewise advocated technological means to increase water efficiency. In particular this needs 'investments in more efficient irrigation systems, use of new technologies (e.g. soil moisture and canopy sensors) to better match irrigation with plant needs, and good agricultural practices', such as conservation tillage, management of soil fertility and water retention capacity, and scheduling of irrigation during night to reduce evaporation (COPA-COGECA, 2007: 4). The basis for improvement is described as follows:

"…. water efficiency measures that provide complementary benefits, such as reduced energy needs or other environmental benefits, will also deliver better results. In many Member States, efforts are being made to increase the water storage capacity of soil under agricultural land use. The modernisation of irrigation systems has steadily progressed and water productivity has also improved considerably (COPA-COGECA, 2013: 3)."

As indicated above, greater water-use efficiency depends on better agricultural practices alongside extra technology. Yet companies generally promote irrigation technology as if it inherently brings all the benefits (interview, COPA-COGECA, 08.07.13). Improperly managed 'hi-tech' systems can be as wasteful and unproductive as poorly managed traditional systems (Perry et al., 2009). When incorrectly applied, irrigation technology 'can cause losses arising on investments made by farmers, thus decreasing the economic water productivity index and the overall sustainability' (Battilani, 2012).

Beyond a problem-diagnosis of inefficiency, moreover, intensive farming practices can degrade soil and water resources, especially through more input-intensive farming in crops such as maize, vegetables, orchard and vine cultivation:

> "Intensive arable production is partly responsible for poor soil structure, soil erosion, loss of soil OM [organic matter] and pollution from fertilisers and pesticides.... The expansion of maize cropping and the move to growing winter cereals in particular have contributed to soil erosion even further (Miller, 2007: 44–45)."

Such harmful practices have been driven and supported by EU policies. In past decades CAP subsidies have tended to favour crops with high water demands, such as maize, thus increasing the risk of water shortages under climate-uncertain conditions (Garcia-Vila and Fereres, 2012). Either as price-support or area-based, CAP subsidies likewise have ensured the profitability of some water-intensive crops such as cotton which otherwise would be phased out under a market-orientated scenario; likewise water-price subsidies.

In some cases, water-price increases have induced farmers to adopt technology and appropriate practices for conserving water (Caswell and Zilberman, 1985). Yet water-pricing policies often have been ineffective means to reduce water demand (Molle and Berkoff, 2007 and Molle, 2008). Farmers experience rising water prices as an extra penalty. Rather than higher water prices, administrative water allocation or re-allocation lowering the supply often has led farmers to adopt water-efficiency practices (Molden et al., 2010). If agricultural water demand is inelastic, then policies which encourage changes in cropping patterns can be more effective than higher prices (Fraiture and Perry, 2007, Iglesias and Blanco, 2008 and Kampas, 2012).

Inelastic water demand results from farmers' perspectives on water benefits. Water-use efficiency (WUE) and water productivity (WP) are often used interchangeably but have different meanings. WUE specifically means the ratio of biomass produced per unit of irrigation water used, i.e. the sum of transpiration by the crop and evaporation from the soil (Sinclair et al., 1984). By contrast, WP means the ratio of above-ground biomass per unit of water transpired by the crop (Steduto, 2007). Both terms have relevance to farmers' economic goals. WUE interests mainly the water districts or management agencies, while WP interests more farmers and research community. WP better speaks to perspectives linking water usage with production levels and economic benefit (interview, COPA-COGECA, 08.07.13).

Yet even WP remains distant from farmers' perspectives. They generally perceive 'irrigation efficiency' as maximising net revenue rather than saving water (Knox et al., 2012). Policies seek to lower water usage, and river basin managers try to allocate limited supplies, yet water-saving is not a priority for most farmers (Luquet et al., 2005). They manage labour and other inputs to get better economic gains (Molden et al., 2010). Towards that economic aim, most growers make irrigation decisions by relying on subjective judgements, based only on their practical experience and observation (Knox et al., 2012). Consequently, there have been limited benefits from irrigation technology, as well documented in the technical literature; the following examples compare various techniques.

For example, mobile-laboratory evaluations compared the distribution uniformity and irrigation efficiency of various irrigation systems in California. Although microirrigation systems are seen as 'efficient technologies', they were performing less well than traditional surface irrigation methods such as furrows and borders. To gain the extra benefits of such technology, most important is adequate system design, alongside proper installation, operation and maintenance, regardless of the irrigation method used (Hanson et al., 1995).

Howell (2003) and Irmak et al. (2011) reported the attainable application efficiencies for different irrigation methods, assuming irrigations are applied to meet the crops' water needs. Microirrigation has the potential to achieve the highest uniformity (90%) in water applied to each plant, yet poor uniformity and application efficiency can result from various causes, e.g., inadequate maintenance, low inlet pressure or pressure fluctuations, emitter clogging and inadequate system design (Hsiao et al., 2007). Consequently, microirrigation technology has on-farm efficiencies varying from 0.7 to 0.95 (Howell, 2003).

As another example, a Spanish study compared various irrigation methods via the annual relative irrigation supply index (ARIS), i.e. a ratio of water applied versus water required. It found a greater efficiency of solid-set and drip than surface irrigation. But average annual figures conceal great variations in water applied to a given crop and irrigation efficiency at farm level, partly for lack of adequate knowledge. A remedy would be 'actions to improve farmers' water management via a combination of irrigation advisory services and policy measures' (Salvador et al., 2011: 586).

Reasons for those Limitations and ways to overcome them

Given the above water-efficiency limitations in applying irrigation technology, the literature has outlined some fundamental reasons. They include the following: irrigation equipment is promoted as if the technology per se brings various benefits, farmers seek to maximise net income rather than water productivity per se, innovative technologies can achieve the full potential benefits only through appropriate technical advice, and farmers lack a knowledge-system for anticipating effects of specific irrigation practices or for retrospectively evaluating their irrigation efficiency.

Although research has developed technical scheduling procedures to improve agricultural water management, these have been little adopted, for many reasons.

> "The one most frequently mentioned by growers is the lack of perceived [financial] benefits relative to their current practices, which they consider adequate. Ease of use and the expenses involved are also important grower considerations (FAO, 2012)."

Technical advice on irrigation scheduling is little used at farm level; at most, it helps retrospectively to evaluate seasonal approaches (ibid.).

One obstacle is inadequate knowledge about proper irrigation levels and scheduling over a growing season. Farmers generally lack adequate assistance to develop and adopt better approaches for environmental sustainability, while also maintaining their financial and social objectives (Pereira et al., 2012: 39). For example, sub-surface moisture sensors can improve knowledge about a crop's need for water. But the technology has limitations, so farmers need technical advice to interpret the measurements; for example, 'soil humidity sensors are still neither easy to handle nor reliable' (WssTP, 2012: 33). Moreover, these sensors are not well adapted to all soil types; their installation and maintenance requires the employment of specialised technical staff. The same is true for the canopy sensors, whose proper application is limited to some crops and during specific growing stages, periods of day and climatic conditions.

Improvements in irrigation practices depend on quantitative knowledge of farmers' current practices in relation to actual and potential crop water use:

> "Any effort to improve water use efficiency needs to start with the assessment of the actual and attainable efficiencies for the given situation, as quantitatively as possible. This information is fundamental for making rational improvements aiming at raising the overall efficiency to the attainable level (Hsiao et al., 2007: 228, 218)."

But such information is rarely available to farmers.

Such difficulties arise for water-management improvements through expert systems. Decision Support Systems (DSS) have aimed to improve crop water use efficiency at farm and water basin scale, but few are widely applied, given the necessary specialised skills. For a DSS to be successful, the key elements have been: giving farmers a simple, timely, user friendly, free-of-charge, informative system helpful to decide how much to irrigate in everyday practice; tailoring the tools for a large number of crops; calculating the irrigation profitability; and assessing the economic benefit, especially its relevance to the next irrigation. Such benefits have been demonstrated by the Irrinet project in Italy's Emilia Romagna (Battilani, 2012). Thus more reliable information systems and expert capacity are necessary to guide farmers in using water more efficiently (Battilani, 2013). This exemplifies the broader need for farmer training and education in order to improve modern irrigation management.

As a way forward in the UK, expert support has been recently linking farmers' responsibility, economic benefits and practical knowledge.

A 'pathway to efficiency' improves the irrigation network, alongside better practices of soil and water management, e.g. by monitoring whether the right amounts are used at the right place and time. 'Using financial criteria for water efficiency rather than an engineering one appears a sensible approach when assessing irrigation performance at the farm level, since any managerial (e.g. scheduling) and operational (e.g. equipment) inefficiencies associated with irrigation are implicitly included in the assessment' (Knox et al., 2012: 3). In particular, 'On-farm water auditing and benchmarking have the potential to provide useful information to farmer decision making, with respect to identifying operational and management changes to improve irrigation system performance and water productivity, and evaluating potential investments in new technology (and advanced practices) or infrastructures' (ibid: 7).

Such approaches have addressed various obstacles to water-efficiency measures. To exploit the full technological potential requires a broader dissemination of their benefits, specific training of farmers, and coupling properly-designed technological solutions with more precise operational practices to benefit farm economic performance (e.g. Tollefson and Wahab, 1994). In particular, advisory-extension services have enhanced irrigation practices which better fulfil potential benefits of irrigation technology (Hergert et al., 1994, Benham et al., 2000, Ahearn et al., 2003, Genius et al., 2014, Parker et al., 2000 and Gold et al., 2013).

Beyond the farm level, greater resource efficiency also depends on shared responsibility among stakeholders, according to the World Business Council for Sustainable Development:

> "Business undoubtedly has many opportunities to increase its eco-efficient performance and thereby to help de-couple use of nature from overall economic growth... Establishing framework conditions which foster innovation and transparency and which allow sharing responsibility among stakeholders will amplify eco-efficiency for the entire economy and deliver progress toward sustainability (WBCSD, 2000: 6–7)."

Analogous issues arise for service-oriented irrigation schemes, designed so that farmers can flexibly obtain water at their convenience, e.g. through on-demand delivery schedules. Here responsibility has institutional complexities. For example, a water users' organisation (WUO) bears largely fixed costs, as well as somewhat variable energy costs from drainage, excess water application, reuse, disposal, etc. If a WUO or water district relies on gravity-fed water conveyance and delivery systems, then its costs do not vary according to water-volume delivery. In such contexts, if farmers decrease water use, then the WUO must increase water prices to recover its fixed costs. Facing higher water prices, farmers may increase groundwater pumping, thus abstracting more water from aquifers, while distancing their individual practices from any group responsibility. Paradoxically, fostering greater water-use efficiency can generate a financial, environmental and institutional problem.

Given those difficulties for water-efficient techniques, their effective adoption depends on several enabling conditions, especially a policy and institutional context aligning incentives of producers, resource managers and society. Significant synergistic effects can emerge when water-efficiency practices are combined with other agronomic practices such as maintaining soil health and fertility, controlling weeds and avoiding diseases (Molden et al., 2010).

Methods and Sources: EcoWater Project

The above issues and earlier questions have been explored through two case studies of service-oriented irrigation schemes within a larger EU-funded research project, EcoWater (see Acknowledgements). It develops a methodology for assessing eco-efficiency at the meso level. The latter is defined as interactions among heterogeneous actors, e.g. between water-service users and providers. As generally understood, eco-efficiency means a ratio between economic advantage and resource burdens, as a basis to evaluate past or potential changes in a system.

The project uses eco-efficiency indicators to evaluate potential innovative practices including technology adoption. The project aims to: assess various options for innovative practices within a specific system; analyse the factors influencing decisions to adopt such practices; and improve understanding of the socio-technical dynamics that influence such decisions.

In the project's two agricultural case studies, farmers and/or their organisations have already invested in water-efficient technology, going beyond state subsidy alone. The irrigation distribution systems were designed for on-demand water delivery. SCADA technology at hydrants allow farmers to abstract water on demand any time and charges them according to a volumetric tiered water pricing. Each case-study area has strong stimuli for farmers to improve water efficiency, yet the full potential benefits of the technology investment were not being realised, for reasons analysed in the next two sections.

SINISTRA OFANTO CASE

Dating from the 1980s, the Sinistra Ofanto irrigation scheme is among the largest multi-cropped irrigated areas in Italy. It is located in south-eastern Foggia province within the Apulia region. Irrigation is crucial for the region's agricultural production and income, but it also generates resource burdens. Nearly 18.5% of Apulia's agricultural area is under irrigation; consequently, irrigated crops have contributed 69% of the total value of regional agricultural production, recently quantified as 3.8bn Euros (Fabiani, 2010). The entire study area is characterised by a high number of small land-holdings with intensive, market-oriented practices. The main crops are vineyards, olives, vegetables and fruit orchards (in descending order). The pedo-climatic conditions are favourable for intensive cropping, but profitable farming is strongly dependent on irrigation, due to the scant rainfall and its uneven distribution across the year.

The Sinistra Ofanto system commands an area of 40,500 ha stretching along the left side of the Ofanto River, of which 38,815 ha are irrigable lands and 28,165 ha are serviced with irrigation distribution. Designed and constructed for pressurised on-demand delivery schedule, the irrigation system is managed by a large water users' organisation (WUO), the *Consorzio per la Bonifica della Capitanata* (CBC, 1984 and Altieri, 1995). The system diverts water from the Ofanto River and supplies it to growers both by gravity and lifting/pumping, ensuring a pressure head of at least 2 bar at each hydrant to enable farmers using micro-irrigation methods.

The system is already equipped with modern technologies to deliver and use water efficiently. From the diversion structure on the Ofanto

River, water is conveyed to the Capacciotti reservoir through concrete-lined canals and pipe conduits, along which the flow regulation devices are downstream-controlled, thus manually or automatically adjusted through calibrated control devices enabling Supervisory Control and Data Acquisition (SCADA). The Capacciotti reservoir supplies seven concrete-lined storage and compensation reservoirs equipped with downstream-control flow regulation devices that adjust inflows and outflows to feed the district's piped distribution networks based on the downstream water demand.

PVC buried pipes comprise the open-branched distribution networks. Each sector's inlet has a control unit, equipped with flow and pressure metering-control devices. Water is supplied to farms on demand by means of multi-users' electronically-fed hydrants that control and regulate the deliveries, as well as the discharges demanded and thus flowing in the pipe distribution network. These technologies installed along the main infrastructure help keeping conveyance and distribution losses within 5–10% of the total water abstracted from the Ofanto River, as reported by the WUO's engineering staff.

Although the main water supply is surface water, during recurrent water shortages farmers pump groundwater from medium-depth (100–150 m) aquifers, especially since the late 1990s (Portoghese et al., 2013). Furthermore, studies found qualitative degradation of groundwater resources, most likely resulting from seawater intrusion into the coastal aquifer and to deep percolation of pollutants, such as fertilisers and pesticides, from intensive farming activities. Given the urgent need to assess these processes and to avoid their adverse environmental impacts, what are the prospects for water-efficiency improvements of irrigated agriculture in the Sinistra Ofanto area?

Irrigation Patterns and Resource Burdens

The water users organisation (WUO), Consorzio per la Bonifica della Capitanata (henceforth the CBC), is the main irrigation management agency. It is responsible for all the sequential steps along the agriculture water supply chain, i.e. abstraction, conveyance, storage, distribution and final water delivery to farm gates. Established in 1933 by a national law of public interest, the CBC is by statute a non-profit organisation; it bears all the costs for performing its functions, and these costs are recovered through the water tariffs paid by farmers. The CBC enforces the principle of solidarity among the different service areas. Even though the costs for supplying irrigation water differ significantly among areas supplied by gravity and by pumping, the tariff structure does not make such a distinction. Rather, as a tool to manage water use, water fees vary according to demand: volumetric tiered water tariffs progressively increase with the seasonal cumulative volumes withdrawn by each farmer. This structure is enforced through individual water metering at the delivery points; all farm hydrants are equipped with an electro-mechanical delivery device allowing the supply of water only to authorised users and storing information of each irrigation event. Besides simplifying the network operations, this technology proved to be very useful for accurate monitoring and control of water distribution, and for achieving better understanding of the irrigation management practices followed by farmers, especially through the possibility to retrieve and analyse historical data series (Zaccaria et al., 2013).

As an irrigation service provider, the CBC is composed of irrigation service users, i.e. farmers. In performing its daily activities, the CBC attempts to reconcile objectives which may be in conflict. Its technical and administrative choices aim to achieve high water-distribution efficiency in order to maximise the economic benefit to farmers. It aims to improve water distribution and use—at the farm level, through an effective operation of the delivery network, and at field and crop level through the technical support to growers aiming at improved water management skills (ibid). Technical support to farmers was effective in the 1980s–1990s but has declined in the last decade, due to WUO budget constraints and lower revenues from Italy's farm activities.

Irrigation water demand is driven mainly by farmers' perceptions, by the climatic conditions, and by the economic value of crop yields and production factors. Even beyond periods of water shortage, in some areas farmers pump groundwater in order to avoid the following problems: (i) the restricted-flow demand delivery-schedule that prevents the quick completion of irrigation cycles in medium-large farms, (ii) the restricted-frequency demand-delivery often imposed by the CBC during water shortage periods, (iii) the need to arrange water withdrawals with neighbour farmers supplied by the same hydrants, or (iv) the tiered water fees enforced by the CBC. Also, many farmers still perceive groundwater pumping as somewhat cheaper than water supplied by the CBC, even though the contrary was shown by economic analyses (e.g. Portoghese et al., 2013).

As a more fundamental problem, both the farmers' perception and the CBC's analyses ignore the ecological costs of groundwater degradation and remediation. The CBC accepts no responsibility for water-management practices beyond the farm gate. From the growers' standpoint, groundwater pumping aims to increase and/or stabilise the economic benefits of farming activities. Often farmers combine surface water and groundwater for various reasons such as to maximise crop yields and farm net benefit, or to minimise the seasonal water fees payable to the CBC, or to prevent yield reduction arising from high salinity in the groundwater during peak-demand periods. However, this conjunctive use of surface and groundwater is based solely on farmers' economic and technical considerations, regardless of environmental burdens such as aquifer depletion and degradation. Furthermore, fields close to the river banks are often irrigated by growers with water pumped out the river. In all these situations, return flows may result from run-off through the drainage networks, as well as from percolation through the soil profile, finally reaching the downstream reaches of the river, wetlands or the aquifer.

Farm activities generate various pressures on land and water resources, including quantitative depletion and qualitative degradation, especially biodiversity loss in farmland and in the natural environment. This harm has several sources: (i) intensive farming and tillage practices, (ii) fertilisers and pesticides application on cultivated fields, (iii) water abstraction from the Ofanto River, (iv) return flows of degraded water to downstream wetlands and aquifers, (v) over-drafting of groundwater, (vi) salinity build-up in cultivated soils, (vii) energy consumption for water pumping, and (viii) increased CO2 emissions from the energy usage related to pumping, transport, machinery, etc.

Relative to those ecological problems, much greater impetus for innovative practices comes from recurrent scarcity of water supply and the prospect of even greater future scarcity and uncertainty. Those problems in turn result from high water-demanding crops and from irrigation scheduling practices. Such decisions are often based solely on farmers' perceptions; their systems and practices are not monitored to assess the actual performance and efficiency achievements. No systematic technical support is available to growers for their daily or seasonal irrigation planning and scheduling.

Moreover there is detailed evidence of water-use inefficiency at farm level. According to a study of farmers' irrigation practices in a nearby irrigated area with similar features, there were often mismatches between crops' water demand and irrigation applications on several occasions during the season. Although the overall seasonal applied irrigation depths may match a crop's water demand, farmers often under-irrigate during the early crop stages and over-irrigate during later stages; many choose inadequate timings and application depths. Such inadequate applications may be combined with uneven in-field water distribution, often due to the average low uniformity of irrigation systems—especially when not properly designed, evaluated and maintained; consequently, the farm may have up to 20% lower crop yields and income, along with inefficiencies between 20 and 40% due to excessive water applications (Zaccaria et al., 2010). As the main reason for the mismatch, irrigation scheduling practices are based only on farmers' perceptions and experiences; missing is status monitoring of soil or plant water, use of ET-based irrigation scheduling, or any other quantitative techniques (ibid). This study confirms a general problem of water-inefficient practices, as also found in the wider technical literature (e.g. Hanson et al., 1995, Hanson et al., 1996, Burt, 2004 and Salvador et al., 2011; see Section 1).

Innovative Practices for Stakeholders' Consideration

Resource-efficiency could be enhanced by properly utilising several innovative technologies and practices. As listed in Table 1, several feasible options are already installed and implemented in the Sinistra Ofanto area, i.e. along the water conveyance and distribution system or on some progressive farms, but require either some refinements or significant operational improvements to gain their full economic and environmental benefits.

Table 1. Technologies, impacts and necessary improvements

Technology	Level of application	State of implementation	Action	Impacts	Needs and Improvments
Multi-user electronic hydrants	Distribution network/hydrants	Installed and operated	Supply water to authorised users Control, record and store irrigation events Bill farmers according to used volumes Implement tiered water tariff Programmable to supply water during specific time slots	Simplify water distribution and water billing Reduce WUAoperational cost Responsible water use by farmers Reduce water demand	Data transmission Low-pressure alert High-salinity alert Integration with IAS
Variable-speed pumps	Pumping plant	Recently installed S operated	Adjust their operation dynamically Supply the discharges and pressure heads required by the network	Energy saving Reduce WUO operational costs	Data recording and transmission Accurate system analysis Tuning operational regulation Improve energy-efficiency
Shift to improved irrigation methods (sprinkle to mini-sprinkle and trickle)	Farm irrigation systems/cropped plots	Recently implemented by some farmers	Reduce soil wetting Reduce soil evaporation losses Reduce energy requirements	Reduce water demand Energy saving Reduce irrigation cost to farmers Reduce overall water abstraction	Better visualisation of benefits Technical support and extension Irrigation advisory service Improve application efficiency

Sub-surface drip irrigation (SDI)	Farm irrigation systems/cropped plots	Very limited	Reduce/minimise soil wetting Reduce application losses Increased WP	Reduce water demand Energy saving Reduce irrigation cost to farmers Reduce overall water abstraction	Precision operation Better visualisation of benefits Improve irrigation scheduling Improve application efficiency Technical support and extension Irrigation advisory service
Regulated deficit irrigation (RDI)	Cropped plots	Very limited	Increased WUE Increased WP	Reduce water demand Energy saving Reduce irrigation cost to farmers Reduce overall water abstraction	Soil and plant water status monitoring Precision irrigation Better visualisation of benefits Technical support and extension Irrigation advisory service

Investment costs for the installed technologies were paid by the WUO either from budget surpluses (in years when water demand and delivery were high), or by funds obtained from different sources, and by some individual farmers at the farm level. Their maintenance and operational costs are paid by growers within the regular water fees. The improved irrigation methods, systems and practices are implemented only to a small extent by some farmers.

The first two technologies were recently installed along the distribution networks to ease operation and flow control, and thus increase its efficiency, but they still need further testing and tune-ups to fully exploit their potential. For example, multi-user hydrants could be hooked up with soil moisture sensors or plant–water status sensors to schedule irrigation applications on the basis of needed or user-defined thresholds and frequency. Also, they could be set to cease water withdrawals and irrigation when the pressure-head values at delivery are insufficient to achieve the target level of efficiency and distribution uniformity. Moreover, relevant datasets from the field (on irrigation timings, flow rates, pressure heads, volume and duration) could be automatically transmitted to a central WUO workstation, thus easing the data-gathering necessary for billing farmers, for evaluating the irrigation delivery service provided and growers' irrigation scheduling practices. These further adjustments of the available technologies could also enable the performance evaluation and benchmarking of irrigation at on-farm and project level.

As indicated by those examples, upgraded technologies alone do not ensure the anticipated resource- efficiency improvements. Compared to less technology-intensive systems, the past investments incur extra costs but bring uncertain gains. The full potential benefits depend on adequate management practices, in turn dependent on better expertise, skill-gap assessment, auditing monitoring and evaluation, control systems and extension programmes, including mission-oriented research and outreach.

The above technology options (in Table 1) were presented at the EcoWater project workshop, held in the case study area and attended by diverse stakeholders involved in the Sinistra Ofanto water-system value chain. These included: the River Basin Authority, the WUO, the Apulia Regional Administration, the regional Department of Water Resources, the Ente Irrigazione, ARPA Puglia, CNR-IRSA (Water Research Institute),

University of Bari, etc. Stakeholders, decision-makers and experts were asked to share their visions about water management in the area. The discussion identified several issues, emphasising the need for measures to protect groundwater.

As many comments there highlighted, in the last three decades contradictory interests have generated conflicts over water allocation and use. Surface-water availability for agricultural use has been reduced by several factors: expansion of the command areas, incentivised partly by irrigation systems and by regional policies; greater water demands of the municipal, industrial and tourist sectors; and adverse environmental-climatic changes. Participants mentioned several solutions, in particular: monitoring and control of water use should be strengthened throughout the Apulia region, especially in the study area, in order to avoid aquifer over-exploitation and uncontrolled withdrawals from water courses; deficit irrigation management, already practiced in the study area by some innovative growers, offers a feasible option to enhance water-use efficiency and productivity; and, finally, farmers should share their personal experiences and knowledge in order to help others to use irrigation water in a sustainable and efficient way (reported in Ecowater, 2013).

In addition, CBC (local WUO) representatives emphasised the importance of using non-conventional water sources as an additional supply, especially the re-use of treated wastewater (TWW) as a valid alternative to aquifer exploitation for irrigation purposes. TWW re-use has several barriers: stringent quality criteria set by regulations for water reuse in agriculture; the high cost of treatment and water conveyance from the treatment plants to agricultural areas; and inadequate research dissemination, even disinformation; and reluctance by farmers. Consequently, TWW re-use should be fostered by several means – lowering the water-quality restrictions, installing the latest technologies available on the market for wastewater treatment to reduce costs of treatment plants, and broadly disseminating research findings to growers, especially through extension activities and farmers' advisers – according to the WUO representatives (reported inEcowater, 2013). No stakeholder raised concerns about aquifer characterisation, safe-yield assessment, groundwater recharge or agrochemicals' fate in soil and aquifers, or farm-level performance assessment of agricultural water management practices.

As emerged during the stakeholder discussion, some future visions may be based on doubtful assumptions. In particular the CBC staff and farmers' representatives assumed that Sinistra Ofanto farmers already achieve high irrigation and water-use efficiency, simply on grounds that they use microirrigation methods. From this assumption, there would be little scope or incentive to growers for further improvement in farm-level water management.

From CBC technical reports over at least a decade, no recent information is available on whether the farm irrigation systems are properly designed, installed, operated and maintained—nor on whether irrigations are adequately scheduled and conducted by growers. Likewise the WUO has not recently fostered irrigation systems evaluations regarding the actual application efficiency and distribution uniformity achieved by growers, or WUE/WP for the various crops and irrigation methods, on the basis of quantitative measurements. Agricultural extension activities have been significantly limited in the last decade by budget constraints. So the CBC lacks an empirical basis for its efficiency assumptions about current performance of on-farm water management. In all those ways, the CBC's practical responsibility ends at the farm gate.

Moreover, responsibility for water-supply problems is displaced onto public-sector agencies, especially for additional water supplies via wastewater reuse. This option is simpler and perhaps more profitable for farmers and the CBC, especially if public bodies carry the main financial burden. Water reuse involves complex issues needing careful investigation and management, in particular: different water qualities may be required, contingent on the final water uses. Water that needs to be buffered or stored for seasonal reuse (e.g. groundwater banking or surface storage) may have other quality requirements than water used for a direct agricultural reuse. And there are potential risks to health and environment (WssTP, 2013: 19–20).

MONTE NOVO CASE

In Portugal's Alentejo the Monte Novo irrigation perimeter provides abundant water for a rising number of farmers, served by the larger Alqueva reservoir and irrigation project. Dating from 2009, the scheme set initially low water prices to incentivise irrigation by farmers—but without their involvement in discussing implications for crop choices, irrigation practices, farming profitability and expected income. The highly-discounted water price has incentivised crops with high water demands as well as agrochemical inputs, thus creating pollution problems. As mandated by law, the water price will rise significantly towards full-cost recovery by 2017, thus potentially stimulating practices which need less water. What are the prospects for water efficiency improvements?

Irrigation Patterns and Future price rise

Since the 1960s Portugal has had plans to irrigate relatively arid areas of the Alentejo, especially to increase employment and livelihoods. Spain's 1993 National Hydrologic Plan was proposing to abstract more of the shared water from the adjoining region, partly on grounds that Portugal was not using the water, but Portugal objected. To avoid such trans-border conflicts, the 1998 Albufeira Convention aimed to protect freshwater and groundwater, as well as the sustainable use of shared water resources, in the context of the negotiation process for the Water Framework Directive (Maia, 2000 and Pulwarty and Maia, 2013). Under the Convention, for example, Spain guarantees annual stream flows to Portugal in normal years (Costa, 2003).

The Albufeira Convention facilitated Portugal's Alqueva Multipurpose Project—the Empreendimento de Fins Múltiplos de Alqueva (EFMA). The EU Structural Funds were expected to pay more than half the investment cost (Vergés, 2001). The 1993 EFMA plan combined a reservoir, hydroelectric dam and irrigation networks, thereby helping to justify Portugal's claim on the water. The project aimed to promote the development of a poor, deprived region through irrigated agriculture, electricity production and tourism. Together these were meant to ensure the sustainability of the project in its early phase (Santos et al., 2011).

From the start the EFMA was criticised as a hydrodinosaur—for accepting engineers' assumptions about future water needs, requiring enormous electricity to transport water and so raising the full cost (Costa, 2003). Before EFMA's operation, maize production had greatly risen alongside greater dependence on CAP subsidies for economic viability. The Agriculture Minister hoped that the Alqueva would irrigate industrial crops such as tomato and sugar beet (Vergés, 2001: 7). By using cheap water from the Alqueva dam, some Alentejo farmers were expecting to capture greater payments from the CAP 1st pillar. So they opposed mid-term reforms decoupling such payments from production levels; so did the Portuguese government (Costa, 2003: 26).

According to its promoters, the Alqueva system will foster significant regional development in Alentejo, both in social and economic terms, due to the strategic water reserve and the associated hydraulic infrastructures. The main objective is to guarantee an adequate volume and quality of water to users, especially irrigation water. Further objectives include the energy hydropower production, urban water supply, ecological conservation and recreational uses (EDIA, 2011).

With its geographical topography and high elevation, the Alqueva scheme was an expensive investment (Costa, 2003), especially relative to the small number of farmers who initially joined the scheme. This arrangement resulted in a high unit cost for the water supply. This has high political-economic stakes, given that full-cost recovery of irrigation systems is required by the Water Framework Directive (EC, 2000).

Irrigation delivery service is provided by the public company EDIA (Empresa para o Desenvolvimento das Infraestruturas de Alqueva), the agency responsible for the Alqueva project development and exploitation. Water is abstracted from the Alqueva reservoir and transported through a network of canals and ducts, from primary network to secondary network through hydrants to farmers. Within the larger Alqueva project, the Monte Novo irrigation perimeter is located in Alentejo district, near Évora municipality. The perimeter provides water for irrigation to an area of at least 7800 ha, while the Alqueva Project will have a total 115,000 ha expected capacity by 2015.

In the Guadiana region including Monte Novo, the tertiary sector has 76% of gross value added (GVA) and 67% of employment; the most important sources are tourist accommodation, catering and public services. The agricultural sector has 10% of the region›s GVA and about 15% of the jobs; average farm size is large (55 ha). These characteristics, associated with the high proportion of farmers in a company structure, indicate the region›s high potential to develop agricultural activities, especially with the full operation of the Alqueva project. But its agricultural activities have a very low competitiveness and a low productivity per unit area; incomes are greatly supported by public subsidies, which comprise about 65% of the total gross margin (ARH-Alentejo, 2011).

In the study area, the major irrigated crops are olives, maize, arable crops (mostly pasture) and horticultures (mostly tomatoes). Farm size ranges from less than 50 ha to more than 500 ha. The largest areas belong to important multi-crop farms, such as the ones owned by the Fundação Eugénio de Almeida (FEA), or the olive farms owned by Olivais do Sul (ODS); together they comprise more than 30% of the irrigated area in Monte Novo irrigation perimeter.

Fertilisers and pesticides are applied to all crops, especially tomatoes and maize. Greater irrigation stimulates run-off, leach-outs of fertilisers and pesticides and soil erosion. Together these practices reduce soil organic matter, soil fertility and its capacity to retain water, in turn increasing potential irrigation inefficiencies and need for better water management for the same cultivation level as before. Indeed, land degradation and nutrient depletion significantly constrain opportunities to increase water efficiency (Molden et al., 2010 and IST, 2013).

In 2009 the WUO Associação de Beneficiários de Monte Novo (henceforth ABMN) was established, representing all farmers connected to the Alqueva water distribution system from EDIA. According to the regulatory Decree Law 84/82, the Association has formal recognition by the Ministry of Agriculture, Trade and Fisheries. The ABMN has the role to promote the administration of constructing the hydro-agricultural development. Nonetheless EDIA has carried out the management and operation of the irrigation perimeter, as well as the investments in constructing the irrigation network.

Water prices are differentiated according to the pressure heads provided at farm-gate delivery. Prices are lower in low-pressure blocks, where water is conveyed and distributed by gravity, generally to larger farmers; irrigation inside farms is within farmers' management. Prices are higher in high-pressure blocks, which are supplied by pumping stations; these reduce the need for farmers to bear the costs of farm-level booster pumps. However, in the high-pressure blocks some large farms need their own investment in irrigation equipment and distribution infrastructures. In some cases farmers already invested in important irrigation infrastructures prior to the construction of the Monte Novo irrigation perimeter. In both cases a higher water price is imposed on important farmers who will not necessarily benefit from higher pressure heads at hydrants (Rita and Capelo, 2005). Moreover, the difference in water prices between low-pressure and high-pressure blocks is insufficient for covering the necessary investments for constructing the main irrigation infrastructures within low-pressure farms' areas, as well as the corresponding exploitation costs, mainly energy and maintenance (Froes and Rodrigues, 2005).

Given those high costs, full-cost recovery would impede a full transition from rain-fed to irrigated agriculture. To attract farmers into the scheme, a 2010 law set the initial water price at only 30% of full cost. This initial price shifts more of the total cost onto the supplier, while incentivising maize cultivation, which demands relatively large amount of water but gains a higher market price. According to the ABMN (Monte Novo's WUO), the discount price gives farmers 'a certain hope', but many have difficulties to obtain credit, which is necessary for farm-level investment to electrify the cultivated zones (Silva, 2010).

As originally set by law, the low water price will be increased by 10% per year towards full-cost recovery by 2017 (Nuncio and Arranja, 2011). This legislative requirement has become contentious. At its full-cost price, water would comprise 13% of the total cost of maize production. Through their WUOs, farmers have been lobbying the authorities to delay the price rise. According to the Federação das Associações de Agricultores do Baixo Alentejo (regional federation of farmers' associations), the higher price will not allow farmers to be economically competitive, especially for a large proportion of maize cultivation, estimated as more than 36,000 tonnes aimed at export markets. According to the former Agriculture Minister, however, the higher price will be competitive with nearby regions, as an adequate basis for more farmers to join the scheme (Maneta, 2010 and Rosado, 2012). The future increase could stimulate water-conserving measures and alternative cropping patterns which would be more profitable and/or need less water, while also reducing other resource burdens. But continuous support measures would be necessary for growers to evaluate and implement such options.

Innovative Practices for Stakeholders' Consideration

The on-demand water-delivery system was meant to incentivise water-efficient practices by farmers and thus to reconcile tensions between various objectives of the Alqueva project. Such a strategy put great expectations upon a technological system. At the irrigation-perimeter level the eventual implementation of innovative practices is meant to be supported by the ABMN (WUO), with investment costs if necessary to be covered from budget surpluses. On the other hand, the farm-level water-use improvements are dependent on farmers' own investment or eventually on government subsidies fostering efficiency increases. Some technologies have been applied at farm level, e.g. sub-surface irrigation at FEA and ODS farms, but only to a small extent. Buried drip-irrigation systems need even more careful design, monitoring and operation than surface-drip or other micro-irrigation systems. Beyond the technological investment, greater application would depend on applied research, demonstration activities, outreach, farmers' training and skills development.

The study team assessed various options for lowering input demands (of energy, water, agrochemicals) or for increasing farmers' income per unit water used (Ecowater, 2012), in two main categories as follows. At the irrigation perimeter level, i.e. the distribution network:

- tiered volumetric water tariffs according to actual water use by growers;
- tiered water tariffs according to timing of withdrawals and energy costs, e.g. lower rates at night-time; and
- pressure head delivery.

And at the farm level:

- drip irrigation, reducing water evaporation (especially relevant to maize);
- sub-surface drip irrigation, minimising soil evaporation and facilitating mechanical weed-control or conservation tillage or minimum-tillage methods (especially relevant to vineyards);
- super-high density olive orchards;
- variable-irrigation practices, e.g. through regulated deficit irrigation; and

- alternative crops demanding less water.

The study team presented the above options at an April 2012 workshop. It was attended by representatives and experts from relevant local bodies. They included: EDIA, farmers' representatives (FEA and ODS), the ABMN (WUO), ARH-Alentejo, and the Centro Operativo de Tecnologia do Regadio (COTR), an advisory body for scientific research regarding agricultural development.

Stakeholders' comments converged around the following points:

- All the proposed technologies could add value to the Alqueva scheme.
- Farmers are interested in any technological configuration that might increase their profit margins, which are currently low.
- Given the high investment costs of the irrigation scheme, a successful operation is important in order to lower the unit cost of water through access to more growers (Ecowater, 2013).

Discussion focused on the knowledge lacking for farmers to minimise irrigation intensity and to conserve soil resources. In particular participants made these comments:

- Irrigation intensity must remain within the carrying capacity of the soil (infiltration rate and water-holding capacity), especially in order to prevent surface run-off, leach-outs and erosion—significant environmental impacts that must be taken into consideration.
- Without adequate knowledge for such judgements, farmers may intensify resource usage, thus increasing costs, leach-outs and soil erosion.
- Root-zone soil moisture conditions are not measured to identify in-field variability and vulnerable areas, so most farmers base decisions on their past experience and daily observations of farm conditions.
- Although a network of meteorological stations already exist in the area, relevant information is not available for irrigation planning and scheduling purposes; farmers' access is still under development (Ecowater, 2013).

Regarding the above point about the soil's water-capacity, a COTR expert emphasised soil organic carbon (SOC) as an environmental indicator and objective. Greater irrigation flows undermine the soil's capacity to retain water and nutrients, while also increasing GHG emissions. SOC conservation is more difficult for annual arable crops than for permanent pasture. To avoid further SOC loss and to enhance its role, farmers could incorporate crop residues into the soil and/or choose crops which need less irrigation and chemical inputs. Before any technological change, decision-makers should evaluate the implications for SOC, he argued. This issue was not taken up by other stakeholders at the workshop.

The workshop discussion also considered whether the cultivation of organic crops could be an alternative option, along with bio-labelling to gain a higher market price. According to the ODS representative, farmers would use organic cultivation methods if they could be convinced that their profit would increase. But owners of small farms would not be easily convinced (Ecowater, 2013). Indeed, small-scale growers lack an advisory service and systematic support for linking organic methods with higher-value markets. This institutional gap illustrates wider difficulties for farmers adjusting to new challenges and gaining the full potential benefits of the EFMA.

Discussion and conclusions

Innovative irrigation practices can enhance water efficiency, gaining an economic advantage for farmers while also reducing environmental burdens. Water-efficient methods and better irrigation scheduling could also integrate water and nutrient management, thus minimising agrochemical runoff and leaching problems. To help fulfil this potential, experts have developed various models of water efficiency and environmental benefits. Yet these models are little used for irrigation scheduling; at most, they help retrospectively to evaluate seasonal approaches (FAO, 2012).

In that generally adverse context, our two case studies of irrigation schemes investigated options, stimuli and difficulties to improve water-efficient practices. In both cases, individual farmers and/or their organisations already made significant investments in irrigation technologies, but their implementation has not been systematically evaluated for effectiveness through better knowledge—regarding crops' water use, soil-moisture conditions, irrigation scheduling techniques, crops' yield response to different irrigation management strategies, etc.

Farmers there have no formal responsibility to demonstrate efficient water use (unlike in the UK; Knox et al., 2012). Consequently, farmers pay higher water prices yet do not obtain the full potential benefits through water-efficient practices. Reasons why have been investigated by exploring perspectives of water users organisations (WUOs) and relevant agencies, especially through multi-stakeholder workshops.

The two cases have strong stimuli for more water-efficient practices. In the Sinistra Ofanto case, a main stimulus is the prospect of greater water scarcity and erratic supply, leading some farmers to rely on groundwater, which may progressively become more scarce, over-drafted, more expensive due to higher pumping costs, degraded and/or saline. In the Monte Novo case, a water-abundant system, a main stimulus is the statutory rise in water price towards full-cost recovery; this would weaken the commercial viability of some crops with high water demand, such as current maize cultivation.

Anticipating such future difficulties, stakeholders have discussed options for better practices. In each area, stakeholder representatives attended our project workshop on how farmers' economic advantage could be combined with environmental benefits. Despite the stimuli for more water-efficient practices, however, actors' responsibility has been somewhat displaced onto hypothetical prospects. In particular, for the two areas:

At the Sinistra Ofanto workshop, stakeholder representatives emphasised problems of water scarcity and aquifer salinity. Their comments suggested various possible measures such as region-wide control and regulation, farm-level practices and farmers' knowledge-exchange systems. At the same time, however, the WUO displaced responsibility onto the prospect of additional supply from wastewater reuse; this would depend on significant institutional, policy and technical changes, with an uncertain timescale and cost.

At the Monte Novo workshop, stakeholder representatives expressed interest in farm-level technological improvements which could improve water efficiency. Yet WUO representatives also emphasised the need to maintain a low unit cost of water. Moreover, the WUO publicly advocates a longer-term low water pricing, which would depend on a legislative change contradicting the country's EU obligations and the original mandate for the irrigation scheme.

In both cases responsibility is displaced, thus complementing WUOs' and farmers' assumption that their agricultural water management practices already have adequate efficiency. This attitude arises by default—from seeing no systematic means or incentive for more water-efficient practices, beyond the modern technologies already adopted. Experts generally have sought to overcome such obstacle through better models of evapotranspiration and efficient water use (e.g. Pereira et al., 2012). Yet such models cannot reconcile stakeholders' different understandings of irrigation efficiency: farmers seek to maximise their economic benefits, while water experts or agencies seek to conserve water (Knox et al., 2012).

Therefore a solution would need a continuous knowledge-exchange with three components:

- expert scientific knowledge of crops' water needs, their yield-response to water and the actual on-farm versus attainable efficiency (Hsiao et al., 2007);
- links between farmers' perspectives, innovative practices and their income benefits (Knox et al., 2012); and
- means to lower resource burdens from inputs and pollutants (Miller, 2007).

Relative to such a solution, our two case-study areas have a nominal extension service which lacks the necessary resources, expertise and operational capability in agricultural water management; these services have no external auditing. A knowledge-exchange system does not exist in those areas, so farmers may miss the full potential benefits of the technological investment already adopted. Under current circumstances, irrigation practices will maintain the unknown efficiency level; along with their WUOs, farmers will have weak capacities to make extra efforts for more water-efficient practices, as well as weak incentives to pay for any further investments.

An adequate knowledge-exchange system would depend on greater institutional responsibility for water-related policies and strategies, for example, through an extension service or a functional equivalent through a WUO. This in turn would provide an enabling condition and incentive for all relevant stakeholders (especially WUOs) to share greater responsibility for agricultural water management across the entire water-supply chain, including farm-level practices as well as drainage and leach-out management. On this basis, more water-efficient practices could combine wider environmental benefits with economic advantage for farmers.

ACKNOWLEDGEMENTS

This work was supported by the European Commission through the project, 'EcoWater: Meso-level eco-efficiency indicators to assess technologies & their uptake in water use sectors', a collaborative research project of the 7th Framework Programme, grant agreement no. 282882, coordinated by the National Technical University of Athens (NTUA) during 2011–2014, http://environ.chemeng.ntua.gr/ecowater/. We would like to thank two anonymous referees of this journal for helpful comments.

REFERENCES

1. Ahearn et al. 2003, M. Ahearn, J. Yee, J. Bottom Regional trends in extension system resources. Agricultural Information Bulletin No. 781U.S. Department of Agriculture, Economic Research Service, Washington, DC (2003).
2. Altieri 1995, S. Altieri The Sinistra Ofanto irrigation scheme: management and maintenance problems. L.S. Pereira (Ed.), Bonifica, vols. 1–2Bastogi Editrice Italiana, Foggia (1995), pp. 40–47.
3. ARH-Alentejo 2011, ARH-Alentejo Planos de Gestão das Bacias Hidrográficas–Região Hidrográfica 7–Relatório técnico para efeitos de Participação Pública, Nemus–Ecossistema–Agro.ges [River basin management plan—Guadiana Technical report for public participation] Administração da Região Hidrográfica (ARH) do Alentejo (2011).
4. Battilani 2012, A. Battilani Sustainable Knowledge-based Irrigation Management: The IRRINET package (2012) 〈http://ec.europa.eu/environment/archives/greenweek2012/adriano-battilani.html〉 , 〈http://ec.europa.eu/environment/archives/greenweek2012/sites/default/files/3-9_battilani.pdf〉 .
5. Battilani 2013, A. Battilani Approaches to Improve the Productivity of Water Use in Irrigation [with diagrams] (2013) 〈http://www.

wsstp.eu/files/WSSTPX0001/II_2_%20Adriano%20Battilani.pdf〉.

6. Benham et al. 2000, B.L. Benham, G.M. Linder, M.D. Hinrichs Improving irrigation management in Nebraska's central Platte valley J. Nat. Res. Life Sci. Educ., 29 (2000), pp. 111–115.
7. Burt 2004, C. Burt Rapid field evaluation of drip and microspray distribution uniformity Irrig. Drain. Syst., 18 (2004), pp. 275–297.
8. Caswell and Zilberman, 1985, M. Caswell, D. Zilberman The choices of irrigation technologies in California Am. J. Agric. Econ., 67 (2) (1985), pp. 224–234.
9. CBC 1984, CBC Consorzio di Bonifica per la Capitanata, Cinquant'anni di bonifica nel Tavoliere Bastogi Editrice, Foggia (1984).
10. CEC 2008, CEC Follow up Communication on Water Scarcity and Droughts in the European Union COM (2007) 414 final, [SEC(2008) 3069], Commission of the European Communities, Brussels (December, 2008).
11. COPA-COGECA, 2007, COPA-COGECA COPA-COGECA Position on Water Scarcity and Drought COPA (Committee of Professional Agricultural Organisations), General Confederation of Agricultural Cooperatives (COGECA), Brussels (2007) 〈http://www.copa-cogeca.be/Menu.aspx〉.
12. COPA-COGECA, 2013, COPA-COGECA Water and Agriculture under a Changing Climate COPA-COGECA (2013).
13. Costa 2003, L. Costa A Portuguese Hydrodinosaur: The Hidden Costs of the Alqueva Dam (2003) 〈http://www.hydrologie.org/hydrodinosaures/download/portugal_ang.pdf〉.
14. EC, 2000, EC Directive 2000/60/EC of the European Parliament and of the Council of 23 October 2000 Establishing a Framework for Community Action in the Field of Water Policy EC (2000).
15. Ecowater 2012, Ecowater Deliverable 1.2: Value Chain Mapping of the Agricultural Water Systems Ecowater (2012) EcoWater project, 〈http://environ.chemeng.ntua.gr/ecoWater/Default.aspx?t=238〉.
16. Ecowater, 2013, Ecowater Deliverable 6.1: Synthesis report from the 1st Round of Case Study Events Ecowater (2013) 〈http://environ.chemeng.ntua.gr/ecoWater/Default.aspx?t=238〉.

17. EDIA 2011, EDIA Portal de mapas [Map Portal] EDIA (2011) 〈http:\\www.edia.pt〉 (last visit in December 2011).
18. EIO 2011, EIO Water Innovation: How Eco-innovation can Contribute to the Sustainability of Europe's Water Resources Eco Innovation Observatory (2011).
19. EP 2008 Addressing the challenge of water scarcity and droughts in the European Union. European Parliament resolution of 9 October 2008 on addressing the challenge of water scarcity and droughts in the European Union (2008/2074(INI).
20. Fabiani, 2010, Fabiani, S. (ed.), 2010. Aspetti economici dell'agricoltura irrigua in Puglia. Istituto Nazionale di Economia Agraria (INEA), http://dspace.inea.it/bitstream/inea/520/1/SE5-1164.pdf.
21. FAO 2012, FAO Irrigation and Drainage Paper 66: Crop Yield Response to Water (2012) 〈http://www.fao.org/nr/water/infores.html〉 .
22. Faurès and Svendsen 2007, J. Faurès, M. Svendsen, H. Turral Reinventing irrigation D. Molden (Ed.), Water for Food, Water for Life: A Comprehensive Assessment of Water Management in Agriculture, Earthscan and International Water Management Institute, London, Colombo (2007) (Chapter 9).
23. Fraiture and Perry 2007, C. Fraiture, C. Perry Why is agricultural water demand unresponsive at low price ranges? F. Molle, J. Berkof (Eds.), Irrigation Water Pricing: The Gap between Theory and Practice, CABI, Wallingford (2007), pp. 94–107.
24. Froes and Rodrigues 2005, J. Froes, P.N. Rodrigues Distribuição de água em baixa pressão no Alqueva Paper in I Congresso da Rega e DrenagemBeja, Portugal (2005) Available in 〈www.cotr.pt/informacao/web2/Papers/48.pdf〉 .
25. Garcia-Vila and Fereres 2012, M. Garcia-Vila, E. Fereres Combining the simulation crop model AquaCrop with an economic model for the optimization of irrigation management at farm level Eur. J. Agron., 36 (2012), pp. 21–31.
26. Genius et al. 2014, M. Genius, P. Koundouri, C. Nauges, V. Tzouvelekas Information transmission in irrigation technology adoption and diffusion: social learning, extension services, and spatial effects Am. J. Agric. Econ., 96 (1) (2014), pp. 328–334.

27. Gold et al. 2013, A.J. Gold, D. Parker, R.M. Waskom, J. Dobrowolski, M. O'Neill, P.M. Groffman, K. Addy Advancing water resource management in agricultural, rural, and urbanizing watersheds: why land-grant universities matter J. Soil Water Conserv., 68 (4) (2013), pp. 337–348.

28. Hanson et al. 1995, B. Hanson, W. Bowers, B. Davidoff Field performance of microirrigation systems Microirrigation for a Changing World, Proceedings of Fifth International Microirrigation Congress, April 2–6, 1995, American Society of Agricultural Engineers, Orlando, FL (1995), pp. 769–774.

29. Hergert et al. 1994, G.W. Hergert, E.J. Penas, R.B. Ferguson, J.E. Cahoon, N.L. Klocke, D.G. Watts, C.A. Shapiro, T.A. Peterson Extension programming for improving nitrogen and irrigation management in Nebraska National Symposium on Protecting Rural America's Water Resources: Partnerships for Pollution Solutions, Washington, DC (1994).

30. Hanson et al. 1996, B. Hanson, W. Bowers, B. Davidoff, D. Kasapligil, A. Carvajal, W. Bendixen Field performance of microirrigation systems Proceedings of the Fifth International Microirrigation Congress, Orlando, FL (1996), pp. 769–774.

31. Howell 2003, T.A. Howell Irrigation efficiency Encyclopedia of Water Science (2003) http://dx.doi.org/10.1081/E-EWS 120010252http://www.researchgate.net/publication/43256707_Irrigation_Efficiency.

32. Hsiao et al. 2007, T.C. Hsiao, P. Steduto, E. Fereres A systematic and quantitative approach to improve water use efficiency in agriculture Irrig. Sci., 25 (2007), pp. 209–231.

33. Iglesias and Blanco 2008, E. Iglesias, M. Blanco New directions in water resources management: the role of water pricing policies Water Resour. Res., 44 (6) (2008), pp. 1–11 http://dx.doi.org/10.1029/2006WR005708.

34. Irmak et al. 2011, Irmak, S., Odhiambo, L.O., Kranz, W.L., Eisenhauer, D.E. 2011. Irrigation Efficiency and Uniformity, and Crop Water Use Efficiency, Lincoln, NB: University of Nebraska–Lincoln Extension, http://extension.unl.edu/publications.

35. IST 2013, Avaliação Ambiental Estratégica dos novos investimentos EFMA relativos à Rede Secundária de Rega, no âmbito do INALENTEJO [Strategic Environmental Assessment

of new EFMA›s Investments] Relatório preliminar: Relatório AmbientalInstituto Superior Técnico (IST), Lisboa (2013) Available online at 〈http://www.inalentejo.qren.pt/images/stories/file980.pdf〉 .

36. Kampas 2012, A. Kampas, A. Petsakos, S. Rozakis Price-induced irrigation water saving: unravelling conflicts and synergies between European agricultural and water policies for a Greek water district Agric. Syst., 113 (1) (2012), pp. 28–38.
37. Knox et al. 2012, J.W. Knox, M.G. Kay, E.K. Weatherhead Water regulation, crop production and agricultural water management—understanding farmer perspectives on irrigation efficiency Agric. Water Manage., 108 (1) (2012), pp. 3–8.
38. Luquet et al. 2005, D. Luquet, A. Vidal, M. Smith, J. Dauzatd More crop per drop: how to make it acceptable for farmers? Agric. Water Manage., 76 (2) (2005), pp. 108–119.
39. Maia 2000, R. Maia Portuguese–Spanish river basins: bilateral agreements' evolution and context Water Sci. Technol., 42 (2000), pp. 227–233.
40. Maneta 2010, L. Maneta Preço da água de Alqueva não convence agricultores Diario de Noticias (abril, 2010) 〈http://www.dn.pt/inicio/economia/interior.aspx?content_id=1540958〉 .
41. Miller 2007, A. Miller Environmentally Sustainable Agri-Food Production Environmental Sustainability Knowledge Transfer Network (ESKTN) (2007).
42. Molle and Berkoff 2007, F. Molle, J. Berkoff Water pricing in irrigation: the lifetime of an idea F. Molle, J. Berkoff (Eds.), Irrigation Water Pricing: The Gap between Theory and Practice, CABI, Wallingford (2007), pp. 1–20.
43. Molle 2008, F. Molle Can water pricing policies regulate irrigation use? 13th World Water Congress, September, Montpellier, France (2008).
44. Molden et al. 2010, D. Molden, T. Oweis, P. Steduto, P. Bindraban, M.A. Hanjra, J. Kijne Improving agricultural water productivity: between optimism and caution Agric. Water Manage., 97 (2010), pp. 528–535.
45. Nuncio and Arranja 2011, J. Nuncio, C. Arranja Gestão dos aproveitamentos hidroagrícolas de Alqueva Jornadas Tecnicas

APRH (2011) ⟨http://sir.dgadr.pt/conteudos/jornadas_aph/apresentacoes/s7/11.pdf⟩ .

46. Parker et al. 2000, D. Parker, D. Cohen, D. Osgood, D. Zilberman Publicly funded weather database benefits users statewide Calif. Agric., May–June (2000), pp. 21–25.
47. Pereira et al. 2012, L.S. Pereira, I. Cordery, I. Iacovides Improved indicators of water use performance and productivity for sustainable water conservation and saving Agric. Water Manage., 108 (2012), pp. 39–51.
48. Perry et al. 2009, C. Perry, P. Steduto, R.G. Allen, C.M. Burt Increasing productivity in irrigated agriculture: agronomic constraints and hydrological realities Agric. Water Manage., 96 (2009), pp. 1517–1524.
49. Portoghese et al. 2013, I. Portoghese, D. D'Agostino, R. Giordano, A. Scardigno, C. Apollonio, M. Vurro An integrated modelling tool to evaluate the acceptability of irrigation constraint measures for groundwater protection Environ. Modell. Softw., 16 (2013), pp. 90–103.
50. Pulwarty and Maia 2013, P. Pulwarty, R. Maia Adaptation challenges in complex river basins around the world Eighth International Conference of EWRA, Water Resources Management in an Interdisciplinary and Changing Context, 26–29 June, Porto, Portugal (2013).
51. Rita and Capelo 2005, C. Rita, O. Capelo Distribuição em alta pressão vs. distribuição em baixa pressão num bloco de rega do E. F. M. Alqueva (Monte Novo) Paper in I Congresso da Rega e Drenagem, Beja, Portugal (2005) ⟨Available in www.cotr.pt/informacao/web2/Papers/38.pdf⟩ .
52. Rosado 2012, L. Rosado A cultura do milho na Fundação Eugénio de Almeida VIII Congresso Nacional do Milho, Lisboa (2012).
53. Salvador et al. 2011, R. Salvador, A. Martínez-Cob, J. Cavero, E. Playán Seasonal on-farm irrigation performance in the Ebro basin (Spain): crops and irrigation systems Agric. Water Manage., 98 (2011), pp. 577–587.
54. Santos et al. 2011, Santos, J.F., Gomes, J.C., Fialho, I., 2011. Contributo para a discussão das condições de sustentabilidade da component hidroagrícola do EFMA, Jornadas Técnicas APRH,

Lisboa, LNEC, 12-15 Outubro, http://sir.dgadr.pt/conteudos/jornadas_aph/apresentacoes/s2/32.pdf, Technical Days index at http://sir.dgadr.pt/jornadasaprh.

55. Silva 2010, Silva, A.M., 2010. Alqueva: água baixa 70% para agricultores, Correo da Manhã, 11 abril, http://www.cmjornal.xl.pt/detalhe/noticias/nacional/economia/agua-baixa-70-para-agricultores, http://boasnoticias.sapo.pt/mobile/noticias.php?id=771.

56. Sinclair et al., 1984, T.R. Sinclair, C.B. Tanner, J.M. Bennett Water-use efficiency in crop production BioScience, 34 (1) (1984), pp. 36–40.

57. Steduto, 2007, P. Steduto On the conservative behaviour of biomass water productivity Irrig. Sci., 25 (3) (2007), pp. 189–207.

58. Tollefson and Wahab, 1994, L.C. Tollefson, M.N.J. Wahab Better Research-Extension-Farmer Interaction can Improve the Impact of Irrigation Scheduling Techniques World Bank, Washington, DC (1994).

59. Vergés, 2001, J.C. Vergés The Alqueva Irrigation Agreement (2001) (Study subcontracted to Modern Consulter S.L. by Research Contractor WRc plc, for the Environment Agency, UK).

60. WBCSD, 2000 Eco-efficiency Creating More Value with Less ImpactWorld Business Council for Sustainable Development (2000) ⟨http://www.wbcsd.org/web/publications/eco_efficiency_creating_more_value.pdf⟩ .

61. WssTP, 2012 Irrigated Agriculture Water Saving Options in Irrigation, Looking for EfficientTechniques, Irrigation Management and Adapted Cropping Practices Water Supply and Sanitation Technology Platform (WssTP) (2012) ⟨http://www.wsstp.eu/files/WSSTPX0001/library/agriculture/SP%20Agriculture%20VI.pdf⟩ .

62. WssTP, 2013 Water Reuse: Research and Technology Development Needs for Water Reuse WssTP (2013) Summary at ⟨http://www.wsstp.eu/files/WSSTPX0001/Water%20Reuse%20Ex.Summary%20online.pdf⟩ .

63. Zaccaria et al., 2010, D. Zaccaria, I. Oueslati, C.M.U. Neale, N. Lamaddalena, M. Vurro, L.S. Pereira Flexible delivery schedules

to improve farm irrigation and reduce pressure on groundwater: a case study in southern Italy Irrig. Sci., 28 (2010), pp. 257–270.

64. Zaccaria et al., 2013, D. Zaccaria, N. Lamaddalena, C.M.U. Neale, G. Merkley Simulation of peak-demand hydrographs in pressurized irrigation delivery systems using a deterministic-stochastic combined model. Part II: Model applications Irrig. Sci., 31 (2013), pp. 193–208.

Chapter 2

Property Rights and Sustainable Irrigation: A Developing Country Perspective

Ruth Meinzen-Dick

International Food Policy Research Institute, 2033 K St NW, Washington, DC 20006, USA

ABSTRACT

While the role of secure property rights contributing to sustainable natural resource management is increasingly recognized, translating that into practice is more challenging, especially in developing countries. This article presents a framework for understanding the role of property rights for effective irrigation systems and then explores the complexity of property rights to land, water, and infrastructure and their underlying institutions. Understanding property rights in practice requires acknowledging legal pluralism—the coexistence of many types and sources of law, which can be used as the basis for claiming rights over the resources. Property rights do not necessarily imply full ownership, but are composed of different bundles of rights that may be held by different claimants—the state, user groups, families, or individuals. These rights are critical for the authority, incentives, and resources for irrigation operation and maintenance. As resources become more scarce, property rights systems need to adapt to reduce conflict and provide incentives for saving water. However, efforts to improve irrigation by changing property rights systems have often failed because they have not recognized the difficulty of transplanting property rights systems from one place to another. Institutional change needs to be seen as an organic process, building on existing norms and practices, rather than as an exercise in social engineering.

INTRODUCTION

Experience with the past 30 years of irrigation has shown that technology alone is not sufficient to ensure productivity gains, let alone sustainability. In many cases, the technologies were not adopted or maintained, or the poor, women, and other marginalized groups were excluded from the benefits of technologies. Appropriate institutions are needed to accompany technologies for sustainable irrigation.

Among these key institutions, property rights play a particularly important role. When resources are abundant, there is little need to define property rights, but as the resource becomes more scarce, users of the resource compete and even come into conflict. In such contexts, there is pressure to define property rights over the resource, to clarify expectations, and assign both rights and duties (Otsuka and Place, 2001 and Young, in press).

While the contribution of secure property rights to sustainable natural resource management is increasingly recognized in calls for policy reforms, translating that into practice is more challenging (Deininger, 2003). This applies to almost all natural resources, including land, forests, and fisheries, but even more so to water, which is a mobile and often fugitive resource, difficult to measure or control, with many different uses and values, both economic and noneconomic (Rogers et al., 1995). Efforts to "improve" irrigation by changing property rights systems have often failed because they have not recognized the difficulties involved in institutional change, especially of property rights.[1]

While water rights, and especially water rights reforms, are complex in any context, the challenges are particularly severe in developing countries.[2] A major reason for this is the frequent lack of state capacity to carry out the reforms as envisaged (as described by Garduno, 2005 for Mexico or Schreiner, 2013 for South Africa). However, customary water rights are also likely to be particularly strong in developing countries, often varying from one context to another, and differing from state law (von Benda-Beckmann et al., 1998). Moreover, as North (1990) notes, institutional change is path dependent: it is inherently shaped by the history of a particular place. This means that transplanting property rights systems from one place to another will not yield the same results, as shown by Solanes (2013) in his analysis of efforts in the 1990s to pass a water law in Peru that was modeled after Chile's legal framework of highly privatized water resources.

Thus while property rights can play a crucial role in sustainable irrigation in developing countries, any reforms to achieve this aim need to be based on a sound understanding of property rights to land, water, and infrastructure and their underlying institutions. Of these, land rights have been dealt with most extensively, both in scholarship and in policy (see, for example, Deininger, 2003 or FAO, 2012). The focus of this article is therefore on rights to water and secondarily irrigation infrastructure. However, rights to all three of these key resources are inextricably intertwined (see Hodgson, 2004).

This article presents a framework for understanding the role of property rights for effective irrigation systems and then explores the complexity of these property rights, with particular reference to conditions in developing countries. Applying the framework helps to identify the situations in which institutional issues of coordination and property rights are likely to be important for sustainable irrigation (and where, conversely, they are less likely to require much attention). Subsequent sections of the article discuss some of the institutional complexities, particularly of property rights, and their implications for programs seeking to improve the sustainability of irrigation systems.

KEY INSTITUTIONS FOR SUSTAINABLE IRRIGATION

Institutions are the "rules of the game" that govern behavior in society (North, 1990:1). Fig. 1 illustrates the importance of two types of key institutions for irrigation and agricultural water management: those that provide coordination and property rights. The vertical axis illustrates the spatial scale of a technology, from an individual plot, through a whole farm, to one that covers several farms, a village, to a region. All approaches that are above the scale of the individual farm require some form of coordination—either by local organizations, the state, or the market. For example, a drip irrigation kit may be adopted by an individual small farmer, by his or her self. Even a well may serve just one farm, but where holdings are very small and tubewells have large capacity, farmers may join together to buy and operate a tubewell, or the state may install and operate it, or one farmer can install it and sell water to neighbors. As we move up the spatial scale, a small check dam may serve a group of farmers. A watershed management program may serve several communities. Canal irrigation systems and reservoirs can serve up to thousands of farmers, and even cross provincial lines. Finally, transboundary river basins cross national boundaries. In each case, some form of coordination is required to govern provision and expropriation: to ensure that the infrastructure is built and maintained, and to allocate and distribute the water among users of the same source, and settle disputes. The greater the spatial scale, the higher the level of coordination that is needed, as indicated by the corresponding arrow on the right side, to balance the spatial arrow on the left. That coordination may be provided by the state or by collective action (or, in some cases, even by the market).

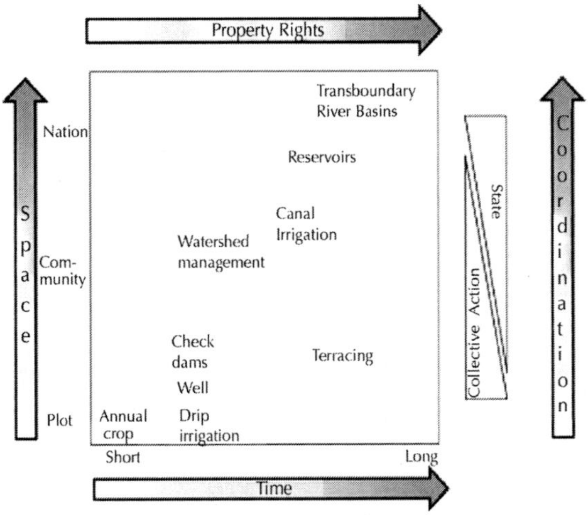

Figure. 1. The role of property rights and coordination institutions for sustainable irrigation.

Similarly, the lower horizontal axis indicates the permanence of a technology or approach, or the time frame between when an investment is made and its returns are realized. The longer the temporal scale, the greater the need for property rights to provide authorization and incentive to make the investment (Knox et al., 2002). Even a tenant or a wife without independent land rights can install a drip kit, but may not be allowed to install a treadle pump or tubewell, because she does not have the decision-making (management) rights over that land, and she may not have the incentive to install and maintain terracing or drainage systems for salinity control. The arrow on the upper horizontal axis indicates that secure property rights become more important as the time frame of irrigation practices increase. Even those with decision-making rights may not have the incentive to make long-term investments unless they have secure tenure, to know that they will benefit from the investment. Although security of tenure is often thought of in terms of rights to the land, people may not be willing to invest in irrigation systems if they do not also have secure rights to the water that will give them the returns. This has been the problem with many irrigation management transfer systems, where farmers were expected to bear the costs, without secure rights to the water from the systems.

While the exact location of any type of irrigation on Fig. 1 would depend on the size of the farms and the scale, as well as the cost/return ratio, of the particular technology, this framework provides a useful starting point to ask which institutions are likely to be critical. As water uses increase, even seemingly "independent" water users will be affected by, and have an impact on, other water uses and users. For example, Asquith (2006) describes how immigration and more intensive land use upstream had impacts downstream in the Los Negros Valley of Bolivia. This increasing interaction between uses within basins calls for better institutional arrangements to coordinate water uses (as described by Asquith, 2006). Property rights are important in this regard because they clarify who can use and manage the land, water, or infrastructure, and what responsibilities they have toward the resource and toward others. Coordination institutions help to set and enforce those rules about the allocation of water and responsibilities. These issues are most apparent in surface water flows and irrigation, but also applies to groundwater irrigation. The following sections provide more details on the institutions for coordination and property rights, with special attention to their implications for sustainable irrigation.

COORDINATION INSTITUTIONS

The example of the tubewell cited above illustrates that coordination functions can be provided by the state (a public tubewell that supplies many farms), collective action (farmer group) or markets (farmer selling, buying or trading water). How well each of those institutions functions will determine whether farmers receive adequate and timely water supplies. For example, in the 1970s and 1980s there were numerous state-run tubewells in India, Bangladesh, and Pakistan that should have allowed economies of scale by using large-scale pumps to irrigate many surrounding farms. These large pumps offer greater potential technical efficiency over small-scale pumps. But these potential technical efficiency gains were often not realized because of poor operational performance of the government agencies running and maintaining the pumps. Instead of government-operated tubewells, a group of smallholder farmers may get together to make the investment in drilling the well, installing a pump, and operating it (collective action coordination), or an individual farmer may install and operate the well, and sell water to neighbors. Shah and Raju (1988) argue that groundwater markets are much more effective than state tubewells in serving farmers' water needs in India. Meinzen-Dick (1996) compared the performance of shared tubewells and groundwater markets in Pakistan, finding that smallholder farmers were quite well served by groundwater markets, but when the water became very scarce, the well owners would deny water to water purchasers, who had little recourse against more powerful well owners.

Which institution is most appropriate depends on the particular conditions—e.g. scale, technical sophistication of the technology and the farmers, cultural factors (social capital, market orientation), and state capacity. In general, the advantages of the state are greatest at the largest scale, and collective action at more localized levels (although there are exceptions). Markets are highly variable in whether they provide effective coordination among smallholders, but seem to come in where machinery is involved, where it is relatively easy to measure water deliveries and exclude those who do not pay (as from a tubewell or piped system), and where users have alternative sources so that there is competition and they are not dependent on one provider.

If group-based (collective action) approaches are selected by programs for water management or technology dissemination, it is important to look beyond the formal rules and membership roles to see whether the group is actually acting collectively: too many water user organizations exist only "on paper" and do not actually work together. Moreover, it is important to consider who is included and excluded from active membership and decision-making. This means asking about women as well as men, land owners and tenants, farmers and other water users (e.g. fishers, livestock keepers, home gardens, domestic users, other enterprises). There may be formal as well as informal barriers to participation, different motivations and returns to be considered. There are indications that organizations with active participation of men and women may be more effective than single-sex organizations in managing resources like water because they draw upon the skills and resources of both, but establishing active mixed organizations is also more difficult than for single-sex organizations, especially where there is high gender segregation in the society (Pandolfelli et al., 2008). All of these factors should be considered in relation to the objectives of the program when identifying the appropriate strategy for establishing or working with groups for irrigation. Unless there is a strong basis of collective action, group-based approaches may require significant investment of time and facilitation to build up the necessary social capital to be effective. Unless the groups are relatively homogeneous with active participation of men and women, further investment may be required to get diverse groups to work together. Where poor literacy and numeracy exist, adult education programs may be needed to support group functioning as well as increase overall human capital for poverty reduction.

Many irrigation management transfer programs that have sought to use farmer organizations for water management (Garces-Restrepo et al., 2007). One clear lesson from this and other irrigation management reforms is that there are no universal formulas or blueprint approaches; rather there is a need for adaptation to local conditions, and as they change over time (Merrey et al., 2007). The triangles on the right-hand side ofFig. 1 are not only an indicator of the broad comparative advantages of state and collective action institutions at each level, but also as a reminder of the overlap of both types of institutions, and the need for them to work together, as in co-management approaches. Instead of the state crowding out collective action, or local groups having to develop all technical capacity on their own, it is preferable to seek collaboration. For example, state agencies may provide technical backstopping, but recognize that farmers have important knowledge of local conditions, or state agencies may facilitate access to external resources or help resolve disputes. This is the premise behind a growing number of catchment forums (as in South Africa) and other multi stake holder platforms. But even this is not straightforward: as Komakech and van der Zaag (2013)indicate in their study of catchment forums in Tanzania that a detailed study of local institutional arrangements is needed as a starting point, with attention to power structures, including whether large farmers, hydropower, or other influential actors will (or will not) participate.

Similarly, there are options for combining market mechanism with user groups and the state. In identifying the potential for water markets, Rosegrant and Binswanger (1994) recognize the importance of user groups to provide coordination among smallholders involved in water trades above the very local level, and overall state regulation. Similar combinations of market, collective action, and state institutions are also seen in payment for environmental services or watershed management programs that compensate upstream users for changing their water use patterns to improve the flow, quality, or timing of water downstream (e.g.Asquith, 2006, Swallow et al., 2002 and Swallow et al., 2006). Because the coordinating institution will need to have at least some control over the infrastructure and even the water itself, the choice of coordinating institution (or institutions) also has strong implications for property rights. The implications of this are discussed in more detail in the following section on property rights.

UNDERSTANDING PROPERTY RIGHTS

There are numerous definitions of property rights, including: "the capacity to call upon the collective to stand behind one's claim to a benefit stream" (Bromley, 1991:15), "the claims, entitlements and related obligations among people regarding the use and disposition of a scarce resource" (Furubotn and Pejovich, 1972:1139–1140), or "the social institutions that define or delimit the range of privileges granted to individuals to specific assets, such as parcels of land or water..." (Libecap, 1998:77).

Key elements in these definitions are that property rights are social institutions: relationships among people with regard to resources. They entail recognized claims and obligations, rights and duties or responsibilities (Commons, 1968). To be valid, rights must be accompanied by a duty of others to respect those rights. The second set of duties or responsibilities are the responsibilities of the right-holder, which may be related to*use* or *provision*: to use the resource in certain ways (e.g. without polluting the water), or to provide for the construction or maintenance of facilities (e.g. irrigation infrastructure).

Ribot and Peluso (2003) distinguish between *access* to resources (e.g. being physically able to get water) and *rights* (entitlements). A tail end farmer may have a right to water, but might not be able to get access because those upstream take too much. A farmer may also have access due to physical location, force, or even stealth, even if he or she does not have a right to the water. Many water users do not have formalized rights to the water they depend on for their livelihoods. Pradhan and Pradhan (2000) provide an example of this distinction between access and rights from the hills of Nepal, in which irrigators from one system would allow farmers from another area to use water under what may be referred to as "tolerated use", as long as the farmers from the second system did not claim rights over that water. However, this access was not enough for the latter farmers, who used a range of strategies to claim rights over that water. Pradhan and Pradhan (2000:201) find that this has broader applicability to many farmer-managed systems:

> "Wherever possible they divert water from new (uncontested) sources, and if this is not possible they try to acquire water from sources used by others by negotiating, disputing, 'stealing' water on the sly, or forcefully acquiring water, often using political or administrative connections. The farmers are not satisfied with just acquiring water; their long-term goal is to legitimize their access to the water source, that is, to establish rights to the water… to get their claims to a share of the water from a water source accepted by other users and competitors."

With regard to irrigation systems, there are three types of property rights that are important: to the land, the infrastructure, and the water. All three of these resources are necessary for productive use of the water, and how rights over these resources are distributed will affect both the amount and distribution of benefits.

Recognizing property rights to water (as well as irrigation infrastructure) often requires going beyond simplistic, legalistic, or externally-imposed assumptions. When discussing water rights with government officials, I have often heard "There are no water rights here" or "The State owns all water." Both of these statements derive from a very narrow understanding of property rights. Conventional images of property rights often derive from images of unencumbered land rights, with fixed boundaries, within which the individual "owner", backed by a state-issued title, can do whatever he (or more rarely, she) chooses with the resource.[3] Those boundaries are seen as rigid and unchanging, dividing people between those on the inside and outside. Certainly these conditions rarely apply to water, particularly in developing countries.

But this view of property rights is flawed, even for land. As noted above, property rights are above all, social institutions, which define relationships among people. They connect more than they divide. Rights are never unencumbered. Instead of "ownership" that implies rights to do whatever one wants to the resource, it is better to think of "bundles of rights" that may be held by different claimants, including individuals, households, water user groups, communities, and the state. Indeed, the inter linkages between these different claimants on the resource can even be conceptualized as a "web of interests" (Arnold, 2002). Furthermore, all rights do not derive from the state: there may be many different sources of property rights. Each of these aspects apply, to some extent, even to immobile property in industrialized countries. But they are especially relevant to property rights in developing countries, and especially to rights over a fluid and often fugitive resource such as water. The following sections provide more detail on legal pluralism and bundles of rights, as they pertain to irrigation systems.

Legal Pluralism

Understanding property rights in practice requires acknowledging legal pluralism—the coexistence of many types and sources of law, which can be used as the basis for claiming rights over the resources or regulations on what can be done with them.[4] This is often thought of in terms of "statutory" and "customary" law, but it is more complex than that. State law itself may be plural, with different – and even inconsistent –.laws issued for different types of water use. Komakech et al. (2012) describe how water use permits are interpreted and enforced differently by various administrative units in Tanzania. Customary law is yet more diverse, differing between communities (however defined) and, because it is often unwritten, even within a community. Moreover, although "custom" is often interpreted as being rigid and unchanging, in fact "customary" law changes, often quite rapidly, to the extent that they may be referred to as "living law" instead (Ehrlich, 1936).

Beyond this conventional dichotomy there are international laws and treaties, such as the Ramsar Convention governing wetlands,[5] or international treaties governing shared watercourses. Each irrigation project generates its own regulations, such as which parts of the command area are authorized to receive water in each season, from which date, and what the water users are required to do (and are prohibited from doing). Water users' associations generate their own rules regarding the manner in which different people may draw water, and what each needs to provide for the system. Because of the fundamental importance of water for life itself, religious laws and rules also create the basis for water rights, as illustrated in Islamic water law

(Faruqui et al., 2001) or the Balinese *subak* irrigation systems, which derive much of the authority for their detailed system of water rights from a series of temples, from the local field canal up to the whole watershed (Lansing, 1991). Even local norms and sayings can become the basis for claiming water rights. For example, in Kenya Onyango et al. (2007) relate how a local proverb that "even the hyena has a right to water" is the articulation of a norm that no one should be denied water, therefore anyone should be allowed to cross even private property to access water.

These different bases for property rights do not exist in isolation. Rather, the various forms of laws or rules overlap, interact, and influence each other. A change in international law may change state law to bring the latter into compliance; a change in national law may end up affecting custom; but also, customary or religious rules may shape statutory law.

However, property rights are only as strong as the underlying institutions – whether state, community, religious, or other – that shape, endorse and enforce the rights. Rights that are not widely known, recognized, or enforced, have little meaning. Meinzen-Dick and Nkonya (2007) describe these overlapping frameworks as "force fields" of varying strength. In the capital city, among a highly educated population that can call upon the courts or other government officials to enforce it, state law is generally stronger than in the hinterland, where even the officials – let alone the general public – are often unaware of the provisions in statutory law. Each type of property rights has associated bearers of the law, such as state authorities, elders, religious leaders, or irrigation system officials. The strength of their authority certainly affects the strength of the property rights that they enforce. However, none of these have the ability to be everywhere and govern every water use or every act to maintain (or damage) an irrigation system. Property rights and obligations also have to be known by those immediately involved, which requires internalizing the rules.

This has important implications for the sustainability of irrigation systems. For any type of property rights to provide security and incentives for careful management of the water or infrastructure, the rules need to be clearly understood as well as enforced. Four of Ostrom's (1990) well-known "design principles" for the governance of irrigation systems relate to this: clearly defined boundaries, monitoring, graduated sanctions, and conflict resolution mechanisms. These principles were developed from a review of many studies of irrigation systems (as well as other types of common property). Cox et al. (2010) validated these principles based on many more such studies and found that it is important not only that the resource and its use are monitored, but that the monitoring be done either by the users themselves or by someone who is accountable to the users.

Bundles of Rights

As noted above, property rights to any resource almost never convey total rights, as is conjured up in notions of "freehold" or "ownership." Rather than holding that up as a standard, it is better to look at the different particular rights that different stakeholders have, which vary enormously, not only between different societies, but also between users within a particular locale. For comparative purposes, these are often identified as *bundles of rights*, but even how these bundles are defined varies. Schlager and Ostrom (1992)define these bundles in terms of *use* rights of access and withdrawal, and *decision-making* (or control) rights of management, exclusion, and alienation. Others, following older Roman tradition, group the bundles as:*usus* rights to *use* the resource; *usufructus* rights to derive income from a resource; *abusus* rights to change the resource; and *alienation* rights to transfer the rights to others (see Eggertsson, 1990, Alchian and Demsetz, 1972 and Scott, 2008). Combining the two classification systems, the following provide examples of the bundles of rights that are particularly relevant for sustainable irrigation.

- *Access* refers to rights for non consumptive use. The "right to ramble" or cross someone's land for a walk is a classic example of such access. Swimming in a reservoir would be another such example. Although access rights should be non consumptive, there may be crowding effects or wearing down of infrastructure if too many people access land or water resources.
- *Withdrawal* refers to rights to take resource units, such as consumptive uses of water or taking produce of the land.
- *Usufructus* rights to earn income from the resource are not included in the Schlager and Ostrom (1992)framework, but can be important where the state charges a water licensing fee. [6]

- *Management* covers rights to make rules and modify the resource, notably setting water delivery patterns or conducting maintenance.
- *Exclusion* refers to rights to determine who can use the resource, and the transfer of those rights. In the case of irrigation, the right to decide who can draw water for irrigation, and whether other users can also access the water, infrastructure, or adjacent land are all crucial exclusion rights.
- *Alienation* refers to the rights to transfer management and exclusion rights to others, such as by sale or lease.

"Ownership" or "complete property rights" are often interpreted as holding all of these bundles of rights. If the term "water rights" conjures up these images, then it is quite understandable that the concept would not seem to apply to water, especially in developing countries.[7] This is especially true if one looks only for formal rights issued by the state. Recognizing this, Hodgson (2013) argues for a focus on *tenure* rather than*rights*:

"A key advantage of couching the issue of water allocation in terms of tenure, rather than simply in terms of water rights, is that it facilitates recognition of the fact that a complex range of different, yet inter-linked, types of relationship exist between people with respect to water resources. More specifically, the question of water tenure goes beyond water rights and water law (Hodgson, 2013:1)."

This is a useful reminder that the focus needs to be on the institutional arrangements, rather than on specific formulations of what is recognized as property rights.

DISCUSSION: PROPERTY RIGHTS AND IRRIGATION MANAGEMENT

Attention to property rights in developing country irrigation systems came to prominence in the 1980s and 1990s in response to concerns with irrigation system maintenance and water scarcity. This section analyzes the policy debates and trends around each of these areas, and the factors that influence security of tenure.

Maintenance

Many irrigation systems have been developed by government agencies, but the performance of these systems has often been far below expectations. With the fiscal crisis of the state in the 1980s, funding for system maintenance was often scarce, causing further deterioration of systems. At the same time, studies of farmer-managed irrigation systems showed that in many cases these performed better than state-managed systems, because the farmers themselves undertook maintenance and irrigation staff were responsible to the farmers (Lam, 1998, Ostrom, 1992, Tang, 1992 and Uphoff, 1986).

Irrigation rehabilitation projects have identified one of the causes of these problems with government-managed irrigation system as being the tendency of the irrigators to feel that the responsibility for maintenance of the systems rested with the government, but the government agencies lacked the resources or capacity to do adequate maintenance. In order to deal with this, many government officials and donor-assisted irrigation programs have called for farmers to develop a "sense of ownership" over the irrigation systems so that they will take care of them (e.g. Vermillion, 1987).[8]Coward (1986) explicitly recognized the link between property rights and management, but focused on the effect of creating and maintaining property together as creating the "social glue" that would strengthen irrigator groups to maintain their irrigation.

Three reasons why property rights are important for maintaining irrigation systems can be summarized as authority, incentives, and resources (Meinzen-Dick, 2002).

- *Authority:* Unless farmers are recognized as having management rights over the infrastructure and water, they do not have the authority to make any changes in the irrigation systems, including maintenance.
- *Incentives:* Property rights, especially *use* and *usufructus* (income) rights, can provide incentives to maintain irrigation systems because they create confidence that the holder of the rights will reap the future benefits of investment and careful maintenance, and bear the losses incurred by misuse of the resources.
- *Resources:* Control over water, land and infrastructure, including *usufructus* rights to earn income from them, can

generate resources (cash or labor) for maintaining irrigation systems.

In most government irrigation systems the state claims the rights of management, exclusion, and earning income, and irrigators only have recognized use rights, often upon payment of an irrigation fee that is then supposed to fund operation and maintenance. However, there is often no direct link between the fees paid by irrigators and the budgets available for the irrigation agency, so the resources are not made available to sustain the irrigation systems. At the same time, farmers lacked the authority to undertake maintenance, and often had little incentive to do so. Recognizing these problems, a number of irrigation management transfer programs have included ceremonies for transferring "ownership" of the infrastructure of surface irrigation systems from the government to farmers' organizations (Vermillion, 1997). However, without rights to the water as well, the infrastructure itself is of little value, and even is a burden to maintain. The water rights are key assets, because they determine both the level and distribution of benefits from irrigation.

This is not to imply that secure property rights will necessarily lead to adequate maintenance. If the holder of the rights (farmers or the state) has the authority but not the technical expertise, if the returns to irrigation are not high enough to provide incentives, or if the costs of maintenance exceed what the usufructus rights can generate, having the rights will not be sufficient. However, the role of property rights merits greater attention than it has received in most discussions of irrigation maintenance or programs to improve irrigation maintenance.

Water Scarcity

As noted above, as water becomes scarce, it brings users into greater competition with each other. Even relatively water-abundant systems may have scarcity at some times, whether during a dry season or drought year, or as water use increases. Clear property rights define allocation rules for who may use how much water, in what ways, at what time and place. By clarifying expectations, these rules can help reduce conflicts. But a further reason for attention to water rights in the context of scarcity is to provide incentives for efficient use and water conservation. Rosegrant and Binswanger (1994) argue that tradable property rights would create incentives for irrigators (and other water users) to conserve water, because what they did not use could be sold to other uses (including to the environment). Thus, rather than requiring some users to give up water, which often causes opposition, this would allow voluntary transfers from less efficient users to those who could use it more productively, with those who give up water receiving payment. It is beyond the scope of this paper to review empirical experiences with water markets,[9] but it is important to note that tradable water rights hinge upon vesting the right to transfer water to others in the water users, rather than in the state. While irrigation management reform programs in many countries have been willing (albeit sometimes reluctantly) to transfer to the users the use, usufruct, management, and exclusion rights, vesting alienation rights in the users has been even more controversial.

Some farmer-managed irrigation systems, as in Nepal and Bali, have developed water rights systems in which proportions of the flows are separable from the land itself. In these cases, farmers have the incentive to use the water efficiently to irrigate as much land as they can with their shares of water. Zekri and Al-Marshudi (2008) describe proportional water rights systems in centuries-old *falaj* systems in Oman that even reserve a share of the water rights to be auctioned to support the system operation and maintenance—thereby combining resources for sustaining the system and tradable water rights in a context of water scarcity.

The question of how water rights systems can adapt to growing scarcity, either through proportional allocations, tradable water rights, or other mechanisms, is likely to become increasingly important with increasing water use and climate change. Sharing ideas from one place to another can assist in adaptation, but copying systems of water rights is not likely to succeed, because it does not account for the variations in the physical and institutional environment. For example, Chilean water law, with tradable water rights, has been promoted as a model for other countries, but Bauer (2004) shows that the Water Code was shaped by the particular history and conditions of Chile, that are not found elsewhere, and even within that context, the Chilean water law has key shortcomings as well as strengths that need to be understood.

Security of Tenure

It is not only the content of the rights that matter, but the security with which they are held. Especially the strength of incentives to sustain irrigation systems depends on how confident right-holders can be that they will hold that right in the future. Key aspects of tenure security include the expected duration of the right and its robustness.

Duration of water rights are complex because of the inherent fluctuations of the water cycle; these are becoming more severe and uncertain with climate change. Thus, it is difficult to know, from one year to another, whether a water right can be realized the next year. Here infrastructure can increase water tenure security, by providing storage (reservoirs or groundwater) to buffer against short-term fluctuations.

Robustness of property rights refer to their ability to withstand challenges. Some of these challenges come from the physical environment (e.g. weather fluctuations or landslides that destroy infrastructure). Other challenges, however, come from the institutional environment. For example, immigrant communities may begin using water from the same source as other right-holders, with some basis for claiming that right. Or the state may decide to extend an irrigation system, creating new claimants, often to the detriment of existing users.

In some cases, robustness of property rights comes from being able to resist such challenges (e.g. by exercising exclusion rights to prevent new claimants from taking water away from the original right-holders). State recognition can play an important role in creating tenure security, especially where the state has enforcement capacity. Yet in many parts of Africa, in particular, even land is held in customary tenure without formal state recognition, and the costs of developing full cadastres and titling programs are very high, often prohibitively so because of the cost of demarcating property, registering it, settling disputes, and then keeping any registries up to date. It is even more expensive and difficult to develop registries of water rights, especially for millions of smallholder farmers, fishers, or other small-scale water users (van Koppenet al., 2007). Hodgson (2013) notes that many minor uses of water are exempt from requirements to register, especially for basic livelihood needs. But this can also create insecurity because it denies those users of state protection. This is particularly an issue in cases where so-called "land grabs" by outside investors have also become "water grabs", taking water away from existing users. For example, Bues and Theesfeld (2012)analyze how large-scale farms in Ethiopia took water away from farmer-managed irrigation systems, because of the outside investors' greater power and ties to the state.

Robustness also comes from being able to adapt to new situations, particularly changes in the hydrologic regime. Formalized state-issued rights may provide less adaptive capacity than systems in which farmers are able to play a greater role in setting and modifying the rules, i.e. when users have management rights.

Both infrastructure and institutions can play a role in increasing security of tenure, especially over water. The role of storage (infrastructure) in helping people respond to droughts has already been mentioned. Pipes that reduce transmission losses can increase confidence that users will receive the water they are entitled to. Meters and information technology (e.g. monitoring others' water uses) can increase users' confidence that if they follow the rules, others will as well because all will be monitored and can be caught if they break the rules. But these technologies do not work on their own. Someone has to operate the reservoirs, read the meters, or provide the information; hence institutional arrangements are needed to ensure the infrastructure delivers.

South Africa's experience with the National Water Act (RSA, 1998) is instructive of the challenges of developing an equitable water rights system that responds to water scarcity, giving due priority to environmental and basic domestic water uses. Despite active consultation in developing the law, implementation has proved a challenge, especially with limitations of state capacity to monitor and enforce (Schreiner, 2013), continuing resistance by powerful water users who stand to lose from the reforms, and limitations on the ability to redress power asymmetries through water alone, without also addressing land rights (Kemerink et al., 2011).

Rather than looking for any prescriptions for property rights reforms, it is more realistic to recognize that changes are likely to be a process of what Cleaver (2001:26) refers to as: "institutional bricolage", a process by which people consciously and unconsciously draw on existing social and cultural arrangements to shape institutions in response to changing situations." An important aspect of this understanding of institutional change—whether in property rights or coordination institutions, is that people themselves are actors, shaping the outcomes, with outsiders (including the state) playing variable roles.

CONCLUSIONS

The development of irrigation systems that serve more than one farm require coordination systems and some form of property rights to at least identify who has what rights to use, manage, and exclude others from the associated land, infrastructure, and water. Even for individual wells and rainwater harvesting, as more and more people start to irrigate, they impact other users in the watershed, and some form of coordination and rules governing water use can help prevent conflicts and give users some assurance that their investments in irrigation will be sustainable.

However, the state is not the only source of coordination and property rights; collective action through customary institutions, newly developed user groups, and even market mechanism can all play a role. Nor do property rights need to include complete "ownership": particular bundles of rights can be held by the state, user groups collectively, or different individual users. Use rights of access and withdrawal, as well as*usufructus* rights to earn income from the resources can provide incentives for investing in and maintaining irrigation systems; management and exclusion rights provide authority to do so; *usufructus* rights provide revenue for operation and maintenance. Alienation rights that are transferrable, either seasonally (rental) or in perpetuity (sale), can provide incentives for those with existing water rights to conserve their use because they can gain income from transferring the unused water to others.

What matters is the security of these rights—the confidence that they provide that others will fulfill their duty to respect those rights. Rights are only as strong as the institutions that stand behind them. Where the state has very little effective capacity – as in remote regions or when the state is weak – then customary rights are often stronger, especially if their legitimacy is reinforced by widely held norms with internalize self-enforcement so they do not have to be explicitly enforced by others. But customary rights alone may not be secure in contexts where the state decides to reallocate water (or land) to other users.

As resources become more scarce, property rights systems need to adapt. Irrigation development projects inherently change property rights, whether explicitly assigning water and land rights, or implicitly by generating rules about who can do what with the water and infrastructure, or simply by creating new "facts on the ground," of where and when water is available, which forces changes in property rights. Many irrigation management transfer projects have included formal ceremonies to confer on the farmers rights over the infrastructure (frequently only the onerous duties to maintain the infrastructure without real rights over its actual use and disposal), but rarely the water. The creation of transferable property rights, as in Chile or Australia, have been discussed but rarely implemented in developing countries (beyond informal markets for groundwater in South Asia).

Recognition of the need for coordinating institutions and property rights for sustainable and efficient management of irrigation is an important step. But identifying the important institutions is relatively easy compared to ensuring that these are in place. Efforts to improve irrigation by changing property rights systems have often failed because they have not recognized the difficulty of transplanting what is seen as clear property rights systems from one place to another. Rapid reforms can be counter-productive. They are unlikely to be fully implemented as planned, and create uncertainty and resentment in the process. Negotiation with stakeholders, looking for ways to compensate those who might lose out, leads to more legitimacy of the reforms.

Institutional change is possible, but it needs to be seen as an organic process, building on existing norms and practices, and recognizing that local people will be active in adapting – or even resisting – external interventions. Analogies of "social engineering" have been misplaced, because they imply a mechanistic approach. What is required is a more nuanced approach, which may involve adaptation to the physical and institutional environment. A good strategy is to start with people's experiences with water—how they access it, what claims they make for their different water uses. This will help to identify the relevant legal frameworks to address. Then an intervention can work to strengthen the claims of poor people for their important water uses. In many cases, water rights become operationalized through user organizations. Ensuring that women, smallholders, livestock keepers, or other poor and marginalized water users are represented in these organizations is an important step to strengthening their water rights, but is often difficult, because of overt resistance from those who do not want to share water rights and decision-making, or because of social challenges of including marginalized groups in local organizations.

Officially-recognized rights also help ensure that their holders have a "seat at the table" in discussions about further water development or land use changes that may impinge on their rights. Joint planning and modeling of water resource development with government agencies and different user groups helps to put this into practice, but it may require strengthening the capacity of both the agencies and the users. There is no single, optimal property right system for irrigations systems—in developing countries or elsewhere. Rather, we need a range of options and the understanding necessary to be able to tailor them to their (ever-changing) physical and institutional context.

ACKNOWLEDGMENTS

This paper draws on the framework of the CGIAR Program on Collective Action and Property Rights (CAPRi), which is part of the CGIAR Research Program on Policies, Institutions and Markets. Special thanks are due to Henning Bjornlund Sarah Wheeler for organizing the conference that led to this special issue, for the invitation to be part of it, and for their insightful comments, encouragement and patience.

REFERENCES

1. Alchian and Demsetz, 1972 A. Alchian, H. Demsetz Production, information costs, and economic organization Am. Econ. Rev., 62 (5) (1972), pp. 777–795.
2. Arnold, 2002 C.A. Arnold The reconstitution of property: property as a web of interests Harvard Environ. Law Rev., 26 (2) (2002), p. 281.
3. Asquith, 2006 N. Asquith Bees and barbed wire on the Bolivian frontier PERC Rep., 24 (4) (2006), pp. 3–6.
4. Bauer, 2004 C.J. Bauer Siren Song: Chilean Water Law as a Model for International Reform Resources for the Future, Washington, DC (2004).

5. Bjornlund and McKay, 2002 H. Bjornlund, J. McKay Aspects of water markets for developing countries: experiences from Australia, Chile and the US Environ. Dev. Econ., 7 (4) (2002), pp. 767–793.
6. Blackstone, 1766 Blackstone Commentaries on the Laws of England (1766). Book II (1766) <http://avalon.law.yale.edu/18th_century/blackstone_bk2ch2.asp> (accessed 11 Aug 2013).
7. Boelens and Hoogendam, 2002 R. Boelens, P. Hoogendam (Eds.), Water Rights and Empowerment, Van Gorcum, Assen, The Netherlands (2002).
8. Bromley, 1991 D. Bromley Environment and Economy: Property and Policy Basil Blackwell, Cambridge, MA (1991).
9. Bruns et al., 2005 B.R. Bruns, C. Ringler, R.S. Meinzen-Dick (Eds.), Water Rights Reform: Lessons for Institutional Design, International Food Policy Research Institute, Washington, DC (2005).
10. Bues and Theesfeld, 2012 A. Bues, I. Theesfeld Water grabbing and the role of power: shifting water governance in the light of agricultural foreign direct investment Water Altern., 5 (2) (2012), pp. 266–283.
11. Chong and Sunding, 2006 H. Chong, D. Sunding Water markets and trading Annu. Rev. Environ. Resour., 31 (2006), pp. 239–264.
12. Cleaver, 2001 F. Cleaver Institutional bricolage, conflict and cooperation in Usangu IDS Bull., 32 (4) (2001), pp. 26–35.
13. Commons, 1968 J.R. Commons Legal Foundations of Capitalism University of Wisconsin Press, Madison, WI (1968).
14. Coward, 1986 E.W. Coward Jr. Direct or indirect alternatives for irrigation investment and the creation of property K.W. Easter (Ed.), Irrigation Investment, Technology and Management Strategies for Development, Westview Press, Boulder, CO (1986).
15. Cox et al., 2010 M. Cox, G. Arnold, S. Villamayor Tomás A review of design principles for community-based natural resource management Ecol. Soc., 15 (4) (2010), p. 38.
16. Deininger, 2003 K. Deininger Land Policies for Growth and Poverty Reduction: A World Bank Policy Research Report Oxford University Press, New York, NY (2003).

17. Easter and Huang, in press Easter K.W., Huang, Q. (Eds.), in press. Water Markets for the 21st Century: What We Have Learned? Springer; Dordrecht, the Netherlands.
18. Easter et al., 1998 K.W. Easter, M.W. Rosegrant, A. Dinar (Eds.), Markets for Water: Potential and Performance, Kluwer, Boston, MA (1998).
19. Eggertsson, 1990T. Eggertsson Economic Behavior and Institutions Cambridge University Press, Cambridge, U.K (1990).
20. Ehrlich, 1936 E. Ehrlich Fundamental Principles of the Sociology of Law Harvard University Press, Cambridge, MA (1936).
21. FAO, 2012 FAO (Food and Agriculture Organization of the United Nations) Voluntary Guidelines on the Responsible Governance of Tenure of Land, Fisheries and Forests in the Context of National Food Security FAO, Rome (2012).
22. Faruqui et al., 2001 N.I. Faruqui, A.K. Biswas, M.J. Bino (Eds.), Water Management in Islam, United Nations University Press and International Development Research Centre, Tokyo (2001).
23. Furubotn and Pejovich, 1972 E.G. Furubotn, S. Pejovich Property rights and economic theory: a survey of recent literature J. Econ. Lit., 10 (4) (1972), pp. 1137–1162.
24. Garces-Restrepo et al., 2007 C. Garces-Restrepo, D. Vermillion, G. Muñoz Irrigation Management Transfer International Water Management Institute and FAO, Rome (2007).
25. Hodgson, 2004 S. Hodgson Land and Water—The Rights Interface FAO Legislative Study 84FAO, Rome (2004).
26. Hodgson, 2013 S. Hodgson Water rights and water tenure Background Paper 3 for FAO Expert Consultation on Water Governance and the Role of Tenure and Rights in Coping with Agricultural Water ScarcityFAO, Rome (2013), pp. 21–23.
27. Ingram and Brown, 1998 H. Ingram, F.L. Brown Commodity and community water values. Experiences from the U.S. Southwest R. Boelens, G. Davila (Eds.), Searching for Equity: Conceptions of Justice and Equity in Peasant Irrigation, Van Gorcum, Assen, The Netherlands (1998), pp. 114–120.
28. Kemerink et al., 2011 J.S. Kemerink, R. Ahlers, P. van der Zaag Contested water rights in post-apartheid South Africa: the struggle

for water at catchment level Water SA, 37 (4) (2011), pp. 585–594.

29. Komakech et al., 2012 H.C. Komakech, M. Condon, P. van der Zaag The role of statutory and local rules in allocating water between large- and small-scale irrigators in an African river catchment Water SA, 38 (1) (2012), pp. 115–126.
30. Komakech and van der Zaag, 2013 H.C. Komakech, P. van der Zaag Polycentrism and pitfalls: the formation of water users forums in the Kikuletwa catchment, Tanzania Water Int., 38 (3) (2013), pp. 231–249.
31. Knox et al., 2002 A. Knox, R. Meinzen-Dick, P. Hazell Property rights, collective action, and technologies for natural resource management: a conceptual framework R.S. Meinzen-Dick, A. Knox, F. Place, B.M. Swallow (Eds.), Innovation in Natural Resource Management: The Role of Property Rights and Collective Action in Developing Countries, Johns Hopkins University Press, Baltimore, MD (2002), pp. 12–44.
32. Lam, 1998 W.F. Lam Governing Irrigation Systems in Nepal: Institutions, Infrastructure, and Collective Action ICS Press, San Francisco, CA (1998).
33. Lansing, 1991J.S. Lansing Priests and Programmers: Technologies of Power in the Engineered Landscape of Bali Princeton University Press, Princeton, NJ (1991).
34. Libecap, 1998 G. Libecap Distributional and political issues in modifying traditional common-property institutions E. Berge, N.C. Stenseth (Eds.), Law and the Governance of Renewable Resources: Studies from Northern Europe and Africa, ICS Press, Oakland CA (1998), pp. 75–92.
35. Meinzen-Dick, 2002 R.S. Meinzen-Dick Property rights and maintenance of irrigation systems Z. Bewässserungswirtschaft (J. Appl. Irrig. Sci.), 37 (2) (2002), pp. 219–236.
36. Meinzen-Dick, 1996 R. Meinzen-Dick Groundwater Markets in Pakistan: Participation and Productivity International Food Policy Research Institute, Washington, DC (1996), p. 105 Research Report.
37. Meinzen-Dick and Nkonya, 2007 R. Meinzen-Dick, L. Nkonya Understanding legal pluralism in water and land rights: lessons

from Africa and Asia B. van Koppen, M. Giordano, J. Butterworth (Eds.), Community-based Water Law and Water Resources Management Reform in Developing Countries, Comprehensive Assessment of Water Management in Agriculture Series 5, CAB International, Wallingford, UK (2007), pp. 12–27.

38. Meinzen-Dick and Pradhan, 2002 R.S. Meinzen-Dick, R. Pradhan Legal pluralism and dynamic property rights CAPRi Working Paper No. 22International Food Policy Research Institute, Washington, DC (2002).

39. Merrey et al., 2007 D.J. Merrey, R.S. Meinzen-Dick, P.P. Mollinga, E. Karar Policy and institutional reform processes for sustainable agricultural water management: the art of the possible D. Molden (Ed.), Water for Food, Water for Life, Earthscan, London (2007), pp. 193–232.

40. North, 1990 D.C. North Institutions, Institutional Change and Economic Performance Cambridge University Press, Cambridge, UK (1990).

41. Onyango et al., 2007 L. Onyango, B. Swallow, J.L. Roy, R. Meinzen-Dick Coping with history and hydrology: how Kenya's settlement and land tenure patterns shape contemporary water rights and gender relations in water B. van Koppen, M. Giordano, J. Butterworth (Eds.), Community-Based Water Law and Water Resources Management Reform in Developing Countries, CABI, London (2007), pp. 173–195.

42. Ostrom, 1990 E. Ostrom Governing the Commons: The Evolution of Institutions for Collective Action Cambridge University Press, New York, NY (1990).

43. Ostrom, 1992 E. Ostrom Crafting Institutions for Self-Governing Irrigation Systems Institute for Contemporary Studies, San Francisco, CA (1992).

44. Otsuka and Place, 2001 K. Otsuka, F. Place Issues and theoretical framework K. Otsuka, F. Place (Eds.), Land Tenure and Natural Resource Management: A Comparative Study of Agrarian Communities in Asia and Africa, Johns Hopkins University Press, Baltimore, MD (2001), pp. 3–33.

45. Pandolfelli et al., 2008 L. Pandolfelli, R. Meinzen-Dick, S. Dohrn Introduction: gender and collective action: motivations, effectiveness and impact J. Int. Dev., 20 (1) (2008), pp. 1–11.

46. Pradhan and Pradhan, 2000 R. Pradhan, U. Pradhan Negotiating access and rights: disputes over rights to an irrigation water source in Nepal B.R. Bruns, R.S. Meinzen-Dick (Eds.), Negotiating Water Rights, Intermediate Technology Press, London (2000), pp. 200–221.
47. Ribot and Peluso, 2003 J. Ribot, N.L. Peluso A theory of access Rural Sociol., 68 (2) (2003), pp. 153–181.
48. Rogers et al., 1995 P. Rogers, R. Bhatia, A. Huber-Lee Water as a social and economic good TAC Background Papers No. 2Global Water Partnership, Stockholm (1995).
49. Rosegrant and Binswanger, 1994 M.W. Rosegrant, H.P. Binswanger Markets in tradable water rights: potential for efficiency gains in developing country water resource allocation World Dev., 22 (11) (1994), pp. 1613–1625.
50. RSA, 1998 RSA (Republic of South Africa), 1998. *National Water Act*. Gov. Gaz. 398, No. 19182, Cape Town, South Africa.
51. Schlager and Ostrom, 1992 E. Schlager, E. Ostrom Property rights regimes and natural resources: a conceptual analysis Land Econ., 68 (3) (1992), pp. 249–262.
52. Schreiner, 2013 B. Schreiner Viewpoint—why has the South African National Water Act been so difficult to implement? Water Altern., 6 (2) (2013), pp. 239–245.
53. Scott, 2008 A. Scott The Evolution of Resource Property Rights Oxford University Press, Oxford, UK (2008).
54. Shah and Raju, 1988 T. Shah, K.V. Raju Ground water markets and small farmer development Econ. Political Weekly, 23 (13) (1988), pp. A23–A28.
55. Solanes, 2013 M. Solanes Viewpoint—The Washington Consensus, Chilean water monopolization and the Peruvian draft water law of the 1990 Water Altern., 6 (2) (2013), pp. 207–217.
56. Swallow et al., 2002 B. Swallow, D.P. Garrity, M. Van Noordwijk The effects of scales, flows and filters on property rights and collective action in watershed management Water Policy, 3 (6) (2002), pp. 457–474.
57. Swallow et al., 2006 B. Swallow, N. Johnson, R. Meinzen-Dick, A. Knox The challenges of inclusive cross-scale collective action in watersheds Water Int., 31 (3) (2006), pp. 361–375.

58. Tang, 1992S. Y. Tang Institutions and Collective Action: Self-governance in Irrigation ICS Press, San Francisco, CA (1992).
59. Uphoff, 1986 N. Uphoff Improving International Irrigation Management with Farmer Participation: Getting the Process Right Westview Press, Boulder CO (1986).
60. van Koppen et al., 2007 B. van Koppen, M. Giordano, J. Butterworth (Eds.), Community-Based Water Law and Water Resources Management Reform in Developing Countries, CABI, London (2007).
61. Vermillion, 1987 D.L. Vermillion Turning over irrigation systems from the government of Indonesia to farmers IIMI Working Paper 2International Irrigation Management Institute, Digana Village, Sri Lanka (1987).
62. Vermillion, 1997 D.L. Vermillion Impacts of irrigation management transfer: a review of the evidence Research Report 11International Irrigation Management Institute, Colombo, Sri Lanka (1997).
63. von Benda-Beckmann et al., 1998 F. von Benda-Beckmann, K. von Benda-Beckmann, J. Spiertz Equity and legal pluralism: taking customary law into account in natural resource policies R. Boelens, G. Davila (Eds.), Searching for Equity: Conceptions of Justice and Equity in Peasant Irrigation, Van Gorcum, Assen, The Netherlands (1998), pp. 57–69.
64. Young, in pressm Young, M., in press. Designing water entitlement regimes for an ever-changing and ever-varying future. Agric. Water Manage, http://dx.doi.org/10.1016/j.agwat.2013.12.002.
65. Zekri and Al-Marshudi, 2008 S. Zekri, A.S. Al-Marshudi A millenarian water rights system and water markets in Oman Water Int., 33 (3) (2008), pp. 350–360.

Chapter 3

Irrigation Management in Coastal Zones to Prevent Soil and Groundwater Salinization

Nicolas Greggio, Pauline Mollema, Marco Antonellini and Giovanni Gabbianelli

[1]Interdepartmental Centre for Environmental Science Research (C.I.R.S.A.), University of Bologna, Ravenna,, Italy

INTRODUCTION

Soil salinization is one of the most widespread soil degradation processes on earth and, worldwide, one billion hectares are affected, mainly in the arid–semiarid regions of Asia, Australia and South America [1]. In Europe, soil salinity has effects on one million hectares mainly in the Mediterranean countries [1]. There are two types of salinization: primary salinization caused by natural events such as sea spray or rock weathering or seepage [2] and secondary salinization that is caused by human activities such as irrigation with salty water, groundwater overexploitation and excessive drainage [1].

Along the Adriatic coast of the Po Plain, freshwater resources are becoming increasingly scarce, because of irrigation and other intense water use, salinization and long periods of drought [3]. Custodio [4] underlines that, especially in southern Europe, the irrigation practices and the water requirements to sustain the coastal tourism industry exhort a strong pressure on water resources.

The impact of groundwater salinization in coastal areas affects both natural vegetation biodiversity and agricultural production, through soil salinization and reduction of freshwater availability for irrigation. Salinization is closely associated with the process of desertification, because salinity may have direct negative effects on crop yields by reducing the ability of plant roots to take up water [5]. The most common salinity effect is a general stunting of plant growth, but not all plants respond in the same way. Grain and corn may reduce their seed production without appreciably plant dimensions reduction. Wheat, barley, cottons seed productions, on the other hand, are decreased less than their vegetative growth [6]. The only agronomical significant criterion for evaluating soil salinization effects is the commercial crop yield decrease [6]. This encourages farmers in certain areas to cultivate salt resistant crops to adapt to high salinity of the soil and groundwater [5]. In our area fresh irrigation water until now was readily available and no salt resistant crops are grown, or other adaptation measures were taken against soil and groundwater salinization of the farmland.

An increase in water and soil salinity within a coastal habitat has direct effects on the natural vegetation and agriculture [5],[7], including decrease in plant species richness, decrease in wetland dry biomass, changes in plant communities, decrease in seedling germination, decrease in surface area of the leaves, decrease in stem length, N2-fixation inhibition, increasing mortality, and indirect effects such as habitat loss for some animal species [8],[9],[10].

In the Ravenna province, more than 68 square kilometers of farmland are at a risk for soil salinization (Figure 3). These are primarily the areas near the Pialasse lagoons and near the rivers open to sea (Figure 1). This agricultural land is at sea level and drained by pumping machines. Many of the pumping machines are located 5 km far from the shoreline and they maintain the water table at 2 meters below sea level during the year, creating hydraulic gradients land inwards and promoting salt-qwater intrusion from the Adriatic Sea [3].

High rates of anthropogenic and natural subsidence and artificial drainage, among others causes, have caused groundwater salinization in the coastal unconfined and semi-confined aquifer near Ravenna, [3],[11] and a subsequent loss of fertile soils and vegetation species richness in the natural areas [7].

In the Ravenna Municipality just about 35000 hectares are agricultural land, of these 30000 are arable land and 5000 are orchards. Irrigation is widely diffuse and more than 50% of farms are equipped for irrigation. Until two decades ago the main source of irrigation water was groundwater from the phreatic and confined aquifers, but the increase of subsidence rates, forced to change the water supply from groundwater to surface water [12],[13], mostly from the Po River and some from the minor rivers flowing from the Apennines.

The phreatic aquifer consists of sandy beach and dune deposits and is unconfined close to the sea and confined by a clay layer further inland (Figures 11and 12). Two dune belts parallel to the coast covered by pine forests form the only topography above sea level. The area in between the dunes belts is used for irrigated and rain-fed farmland, as well as the land behind the older dune belt, 5 km from the coast (Figure 1).

Most of the coastal aquifer near Ravenna is very salty (above 10 g/l) [3][11][14] but, besides some small shallow freshwater lenses at the top of the aquifer, we found freshwater below an irrigated field. In some ways this result is worrying, because irrigation water should not reach the groundwater, so that leaching of nitrates is avoided and waste of resources prevented. Mollema et al. ([15]) concluded from a water budget analysis over the *Quinto* basin watershed (south of the Ravenna urban area) that the amount of water used for irrigation was abundant but only looked at total volumes for the whole basin and did not quantify the contribution of irrigation to groundwater.

The objective of this study is to quantify in detail the effect of irrigation water on the groundwater hydrology with the help of geochemical analysis, geophysical profiling and infiltration measurements. We address, in particular, the mechanism of how excess irrigation water ends up in the aquifer and how that affects the salinity of the groundwater. By doing all this, we assess whether irrigation practices can help to counteract further groundwater and soil salinization.

STUDY AREA

The study area (red rectangle in Figure 1) is part of the *Quinto* basin, the watershed that is confined between two rivers flowing from the Northern Apennines to the Adriatic Sea: the *Uniti* River in the north and the *Bevano* River in the south. The eastern border of the basin is the Adriatic Sea and the western border follows, in part, the *Ronco* River, a tributary of the *Uniti* River.

An east-west profile through the area starting at the shoreline sea includes a beach near the town of*Lido di Dante*, a recent dune belt covered by a coastal pine forest called *Ramazzotti*, planted at the beginning of the 1900's, agricultural land, an older dune belt (paleodunes) covered with a pine forest (*Classe* pine forest), a zone with many active and abandoned gravel quarries that form lakes and more agricultural land (Figures 1 and 5).

Our focus is on a plot of agricultural area (500 by 500 m) in the eastern part of the profile (Figure 2). The area is around a ditch (shown in red). The green dots are wells used to monitor the freshwater lenses, in particular P1S, P2S and P3S are fully penetrating the aquifer (-27m), the others are shallow piezometers (-6m).

The ditch was dug in 1981 as a reservoir for irrigation water that was withdrawn from the *Fiumi Uniti*River. In 2008, the supply for irrigation changed and now it comes through an underground pipe from the *Po* River. The ditch is 500 m long, 5 m wide and 2,5m deep, resulting in a total volume of 6250 m^3; it is completely disconnected from the drainage channels network.

In 2011, the planting schedule was as follows: in the east parcel tomato plants were grown from April to August; Alfa Alfa was contemporaneously grown in the west parcel. Because of the particularly dry 2011 weather conditions, tomato plants needed irrigation water from the early stage of growth onward.

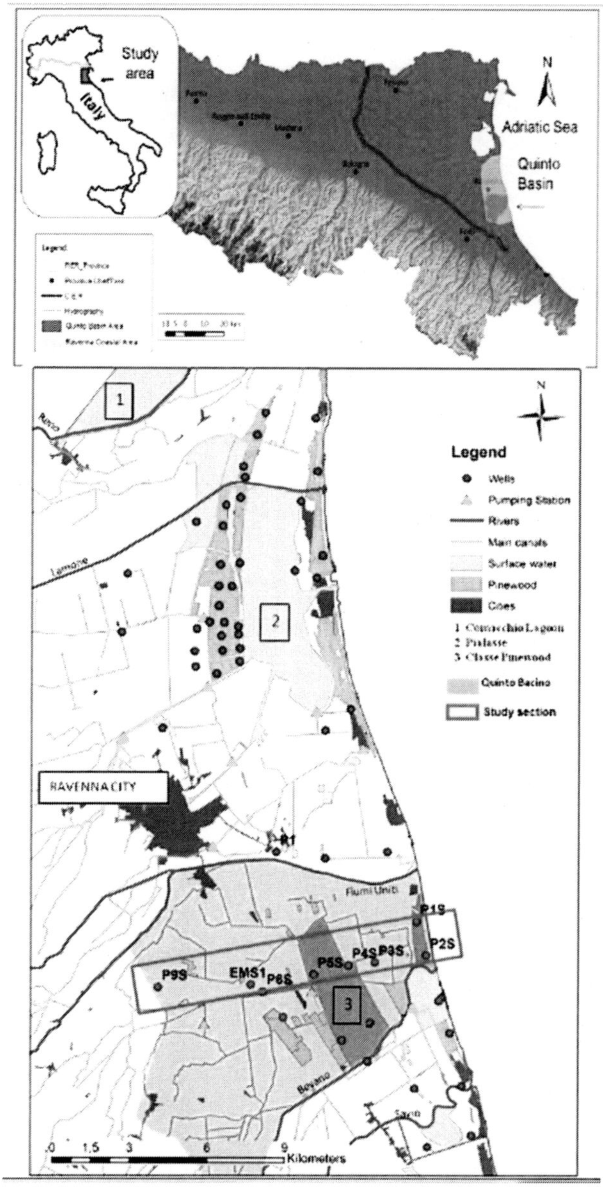

Figure 1. Index map and detail of the Ravenna coastal zone and location of study area (*Quinto* basin); the map reports also monitoring wells, rivers, channels, pumping stations and main environmental features.

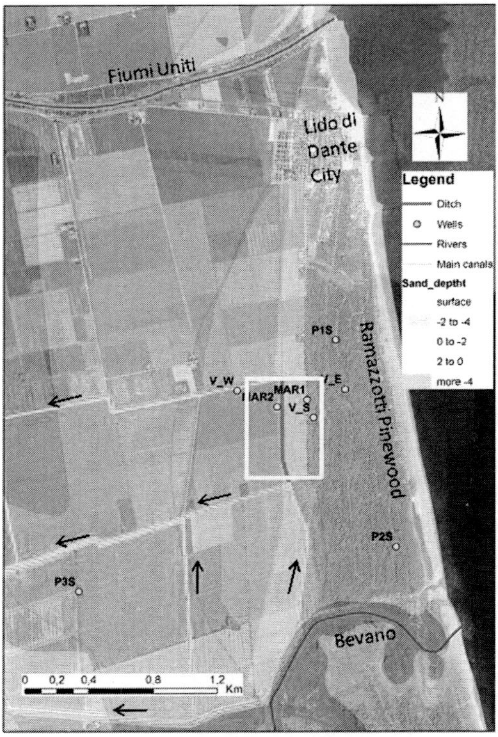

Figure 2. Detail of the study area with thickness of sandy top layer, main drainage channels, flux direction and wells used for water table monitoring.

SOIL TYPE AND LAND USE

In the coastal area, sediments at the surface are mainly sandy (Figure 3). The sand layers are parallel to the coastline and form paleo and recent dunes.

The sandy soils are called *Cerba* soils, and according to the FAO classification [16] for Soil Taxonomy, they are Calcaric Arenosols and Mesic Aquic Ustipsamments [17].

These soil types are mainly used as arable land or horticulture. Due to their high hydraulic conductivity, these sandy soils are most sensitive for salinization [18],[19]. The *Sant'Omobono* soils (Brown in Figure

3) represents river sand deposits; these soils were deposited along recent or paleo-rivers during flooding events. According to the FAO classification [16] for the Soil Taxonomy they are Calcaric Cambisols, Fine Sandy, Mixed, mesic Udifluventic Ustochrepts, [17]. These kinds of soils are cultivated with vineyards and orchards. Because of their sandy texture, also these soils are at risk of salinization. The *Medicina* (*Cataldi*) sediments (Green colors) are soils with fine texture, mainly created when rivers flowed in presently reclaimed areas. FAO classifies them as Eutric Vertisols [16] and for Soil Taxonomy they are Fine, Mixed, Mesic Ustic Endoaquerts. [17]. They are grown with arable crops only and they are not affected by salinization [18],[19],[20].

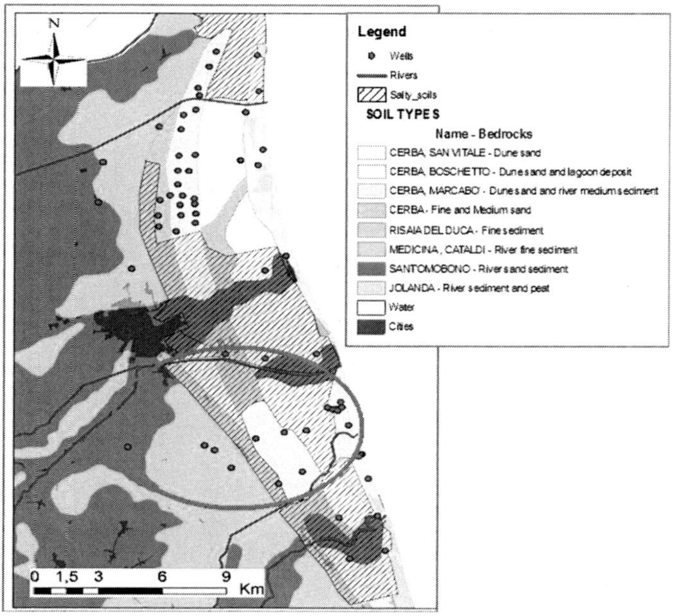

Figure 3. Soil types in the Ravenna area and soils at risk of salinization.

The area studied in detail has an extension of 12 ha consisting of two parcels of land of 6 ha on each side of the ditch. The western parcel lies on a 1-m-thick silty sand top layer (*Cerba Boschetto* soil) that confines the sandy aquifer below, whereas the eastern parcel lies on the unconfined part of the aquifer that consists of beach sand (*Cerba San Vitale* soil; Figures 2 and 4).

The differences in soil and lithology, make the land suitable for different kinds of crops: in the last thirty years, on the sandy soil (eastern parcel), the crops consisted of horticulture (strawberries, tomatoes, potatoes, nursery sugarbeet); on the silty soil instead (western parcel), typical extensive arable crops were grown (barley, wheat, maize, alfa-alfa). At our Mediterranean latitude, horticulture crops need irrigation during all the growth phases [13]; arable crops, on the other hand, do not need irrigation during growth. To provide irrigation water for horticulture, the ditch is kept full of freshwater in the summer period. As a consequence, the water level in the ditch is higher than the water table level in the adjacent aquifer. During the winter, the water level in the irrigation ditch matches the adjacent water table level.

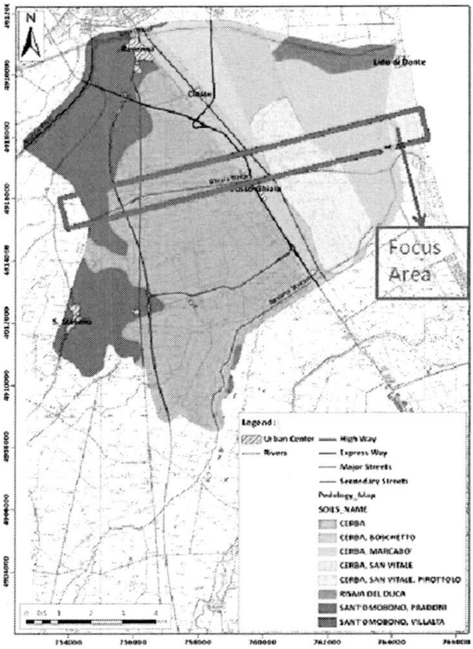

Figure 4. *Quinto* Basin soil types, with study section and focus area.

Most of the land in the coastal zone near Ravenna is used for agriculture but urban and natural areas (pinewood and water surface body) are also widespread (Figure 1). Most of the natural environments are part of the Po Delta Regional Park. Irrigable crops are the most occurring culture in the province of Ravenna. Vineyards and orchards occupy a small percentage of the territory mostly further inland [12],[13].

The *Quinto* Basin covers a total of 10,355 hectares of which the largest part is used for agriculture (66%), 10% is covered by pine forests, 9% is urban area, and 9% is open surface water or wetlands (Figure 5) [15],[21].

The agricultural land is primarily devoted to arable crops as wheat, barley, alfa alfa, sorghum and corn and only 15% is grown with horticultural and orchards. Horticulture represents less than 5% of the*Quinto* basin and it is practiced on sandy soils (*Cerba* soil) (Figure 4). Orchards are grown in the western part, on *Sant'omobono* soils only that, being the deeper soils in the *Quinto* basin, are the most adapt for fruits and grapes, which are sensitive to water logging (Figures 4 and 5). The most common horticultural products cultivated in the *Quinto* basin are tomatoes, potatoes and nurseries of strawberries and sugar beets. The dominant fruit trees are peaches, apples and vine. For this basin, the data from the last general agricultural census shows that the entire surface cultivated with fruits and horticultural crops are equipped with irrigation systems [12].

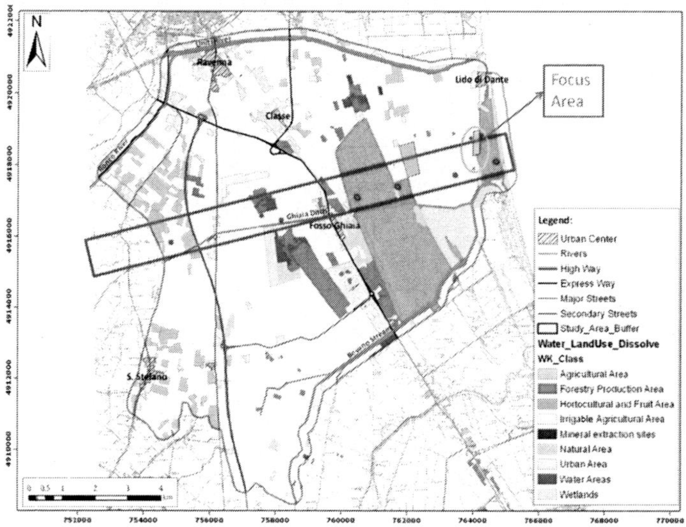

Figure 5. *Quinto* Basin land use map.

TOPOGRAPHY, DRAINAGE AND IRRIGATION

Most of the land near Ravenna is below or just a few centimeters above sea level. Only the recent dune belt and the paleo-dunes are up to 5 m above sea level (Figure 6). These dunes are covered by pine forests (Figure 1). The maximum elevation in the *Quinto* Basin is 5 m and it is measured in the most recent dune belts, whereas the lowest areas are near the gravel quarries 7 km from the coast at 2m below sea level (Figure 6). The farmland around the ditch of our study area is completely flat and lies at a few centimeters above sea level.

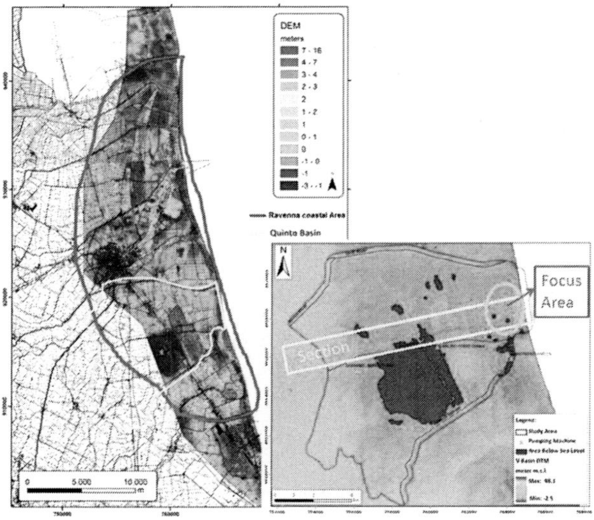

Figure 6. Topography of the coastal zone near Ravenna and area below sea level in the *Quinto* basin.

The whole coastal zone of Ravenna is drained mechanically to make agriculture possible despite the low topography. As a consequence, the areas of natural water infiltration are very small and localized along the dune belts. The entire district is divided into small watersheds defined by small topographic features such as rows of dunes, dikes, and levees [22]. Figure 7 shows the drainage basins in the Ravenna Municipality. Only one basin has a natural hydraulic gradient towards the sea (Figure 7).

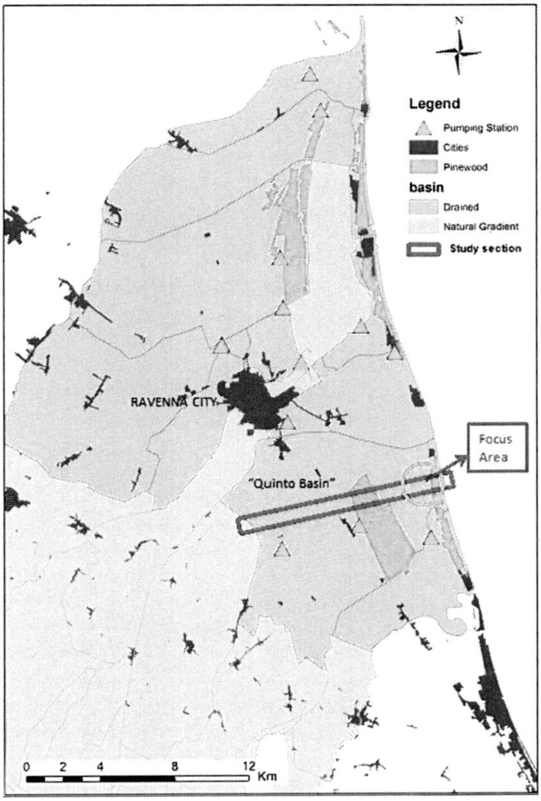

Figure 7. Drainage basins in the Ravenna Municipality. The light green color represents natural drainage basins; the pink color shows the mechanically drained basins.

A dense network of pumping stations and channels was built in the last century with the aim to keep the land from flooding and allow farming [23]. In the *Quinto* basin alone, more than 50 km of drainage channels reroute the water to the pumping stations [24].

The pumping stations are typically located inland, several kilometers from the shoreline (Figure 7). The drainage, therefore, creates a hydraulic gradient directed inland [3],[23]. Our study area is drained by two east-west oriented channels that collect the water between the two dune belts and convey it towards the *Fosso Ghiaia* pumping station (Figure 2).

The pumping stations are activated automatically when the water level in the drainage channels reaches a threshold, regardless of the soil water availability or the weather forecasts. As a consequence, the water table is everywhere below sea level with the exception of the dune belts where the water table may be above sea level in the wet periods of the year. The pumping station in the *Quinto* basin, (*Fosso Ghiaia*) causes a deep withdrawal cone down to -2.5 masl in the western part of the *Classe* pinewood (Figure 8). The flux vectors that were calculated based on water table measurements in the wells ofFigure 1 are all directed inland showing that sea water flows into the aquifer.

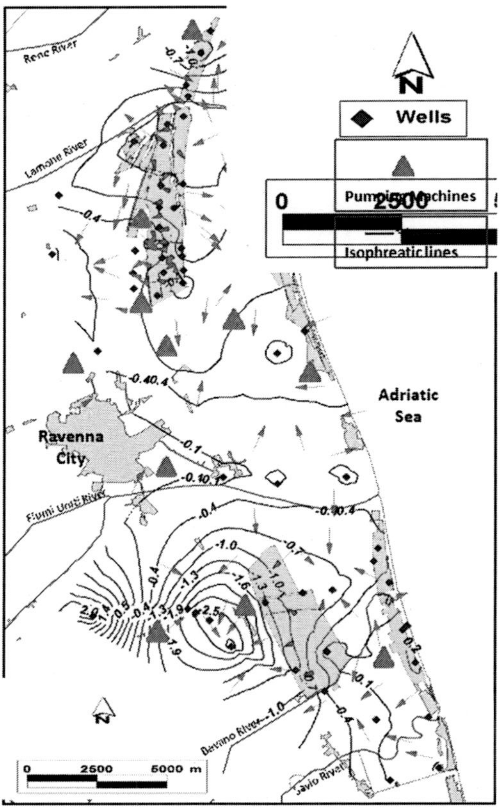

Figure 8. Water table contour map and flux vectors indicating the direction of groundwater flow.

The main source of irrigation water is from river water; five rivers are tapped for irrigation, from north to south they are the *Reno, Lamone, Fiumi Uniti, Bevano* and *Savio* rivers. In the summer period, however, there is not enough water in these rivers and additional irrigation water is taken from the *Po* River by means of the *Canale Emiliano Romagnolo* channel (CER). The water from the CER is brought into the irrigation systems via the same channels that are used for drainage in the winter.

The most common irrigation systems are the sprinkler and the drip irrigation system. The first one is used for horticultural crops with a large volume of water distributed twice a week, whereas the second one is typical for orchards and it uses small volumes of water but with a daily treatment. The sprinkler method, therefore, needs a storage basin for the water (ditch, lake or basin). The drip system instead doesn't need a storage basin [25].

In our focus area, water for irrigation was taken for many years from the *Fiumi Uniti* River. Seawater, however, has encroached along the river and, therefore, it was necessary to change the water supply [3]. Recently, irrigation water comes through a pressurized pipeline directly from the CER (Po river source). During the monitoring time, irrigation on tomato parcel started at an early stage of growth in April 2011 and ended at the harvest time in August 2011. The scheduling of irrigation has been rather regular: approximately twice a week. For each irrigation operation, 20 mm/m^2 were distributed on the tomato plants (Table 1). No irrigation was given to the alfa alfa in the parcel to the west of the ditch.

In the Ravenna Municipality, irrigation and drainage practices are managed by the same authority: the Land Reclamation Consortium for the *Romagna* Region and for both services the end users costs have recently increased [24],[25].

CLIMATE

The climate of Ravenna is Mediterranean with generally warm summers and mild winters [26]. The average temperature is about 14°C varying from -5° to 10°C in the winter period and from 20° to 35° C in the summer months [27]. The average annual rainfall calculated on a long term database (1960-2011) is more than 650 mm and it is usually concentrated in the spring and autumn months.

In five of the last six years, total precipitation has been under the average; in particular 2011 was the driest of the last 23 years, with only 385 mm (Figure 9). Compared with the average monthly rainfall of 1960-2011, in fact, most of the autumn rainfall is absent (Figure 9). The same comparison for temperature shows that in 2011 the temperature was 3-4°C higher than the average of the last 51 years, especially in autumn and winter (Figure 10).

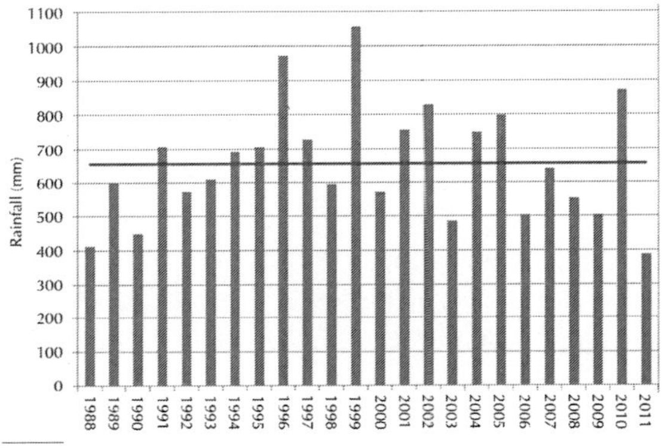

Figure 9. Annual precipitation in the city of Ravenna. Red line represents the long term mean value.

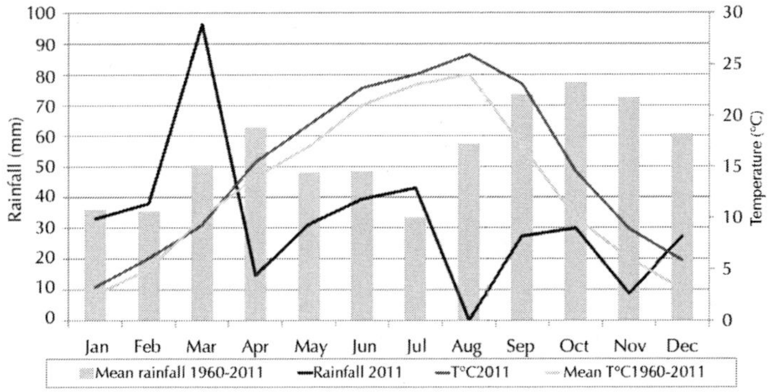

Figure 10. Monthly rainfall and temperature for the Ravenna Area.

One of the climate change scenarios proposed for the Mediterranean area [28] indicates a substantial change in the hydrologic budget by the end of this century. According to Antonellini and Mollema 2010 ([7]), evapotranspiration will increase as a result of an increase in temperature; rainfall will be concentrated in short periods, the winter will become more humid and the summer more dry compared to the present, and there will be an increase in extreme events (floods or droughts). Such a forecast for the coastal zone has a particular significance, because a shortage of fresh water in the area next to the sea will promote saltwater intrusion in the ground water and soil salinization. According to Magnan et al. 2009 ([28]), it is possible to mitigate some of these effects, but it is not possible to avoid them completely; it is, therefore, necessary to start the development of mitigation and adaptation strategies now in order to face these issues.

GEOLOGY AND GEOMORPHOLOGY

The study area is a small portion of the southern part of the wider Po delta plain. The surface aquifer sediments were deposited during the Late Quaternary (Holocene) sea-level fluctuations [29], [30]. During the last glacial maximum (19 kyr ago) the sea-level dropped about 100-120 m in the Adriatic basin and the North Adriatic Sea was became a deep alluvial plain [31] [32]. During the optimum climatic period, about 5500 years ago, the paleo-shoreline was located 25-30 km west of its actual position. From the most landward position of the Holocene coast, a rapid migration of a barrier-lagoon-estuarine system was developed to create the Ravenna alluvial plain, especially from Roman age (IV cent. B.C.) to recent times.

The sedimentation in the study area turned from marine to continental only recently (XIX century). Most of the study area emerged during the last shoreline progradational phase [33], which took place from 1851 to 1894 A.D at an average rate of 20 m/y and in the first half of the past century (4-12 m/y). Nowadays, although the Holocene glacial period with related sea level changes is finished [34], the movements of the coastline still occur: a large part of the coast is under active erosion and only a small part is prograding to sea. This is mostly due to human intervention: the subsidence due to water and gas exploration in combination with a lack of sediments transport by the rivers and the construction of various piers perpendicular to the coast near the harbors and breakwaters parallel to the coast that trap the sediments. These interventions have interrupted the natural process of long shore current and sediment transport [34]. Now beach and dune erosion and shoreline retreat are a major problem along the Adriatic coast [35].

The sandy aquifer consists of two main units: a relatively thick medium-grained sand unit (from 0 to -10 masl) and a lower fine-grained sand unit of small thickness (from -21 to -26 masl). These two sand bodies are separated by a clay-silt and sand-silt continental unit (from -10 m to -21 masl) called*Prodelta* (Figure 11). Lastly, the Flandrian continental silty-clay basement is at a depth varying from -20 m.a.s.l. in the west to -30 m.a.s.l. at the present shoreline [29], [36],[37].

The resulting picture of the aquifer is a wedge shaped dune and beach sand body pinching out in a westerly direction. It is enclosed at the bottom and top by clays and peat deposited in lagoons, marshes and alluvial plains. The only portion where the aquifer is phreatic are the paleodunes and the actual dune belt both covered by pine forests. At the tip of this wedge there are gravel deposits (Figure 11) [29] [38].

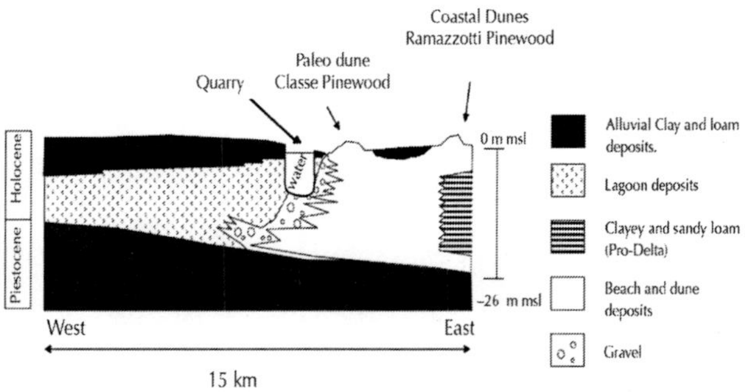

Figure 11. Lithological cross section of the Ravenna aquifer in the study area (modified from [29] [39]).

MONITORING

A groundwater monitoring campaign has been carried out in 8 wells along the section of Figure 1, in December 2010 and it was repeated in

April 2011, with the aim to compare salinity and chemical data in two different seasons. A monthly water table monitoring was performed for the whole transect with increased data density around the ditch, where we installed two shallow wells in February 2011. Overall, 8 wells near the ditch were monitored for water table analysis (Figure 15).

Rainfall was measured with 5 pluviometers located along the section. In the eastern parcel, within the tomato field, we were able to measure both rainfall and irrigation. To assess how much rainfall and irrigation from the surface reaches the groundwater, we buried in the same location a drain-gauge. Geo-electric vertical resistivity surveys (VES) in cross sections normal to the ditch (Figure 12) were carried out in order to determine the depth, shape, and evolution of the fresh water lens [40],[41],[42]. VES surveys were repeated twice: before the tomatoes were planted in February 2011 and during the maximum irrigation period in July 2011. The VES calibration was made using well salinity data, and apparent resistivity data obtained for the area by the process of inversion [43] [44].

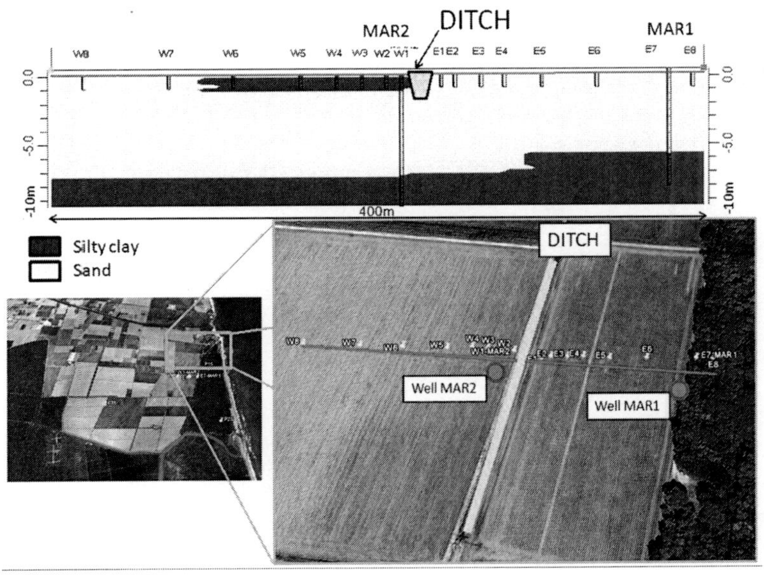

Figure 12. Small parcel, studied in detail, with simplified lithology, irrigation ditch (light blue), wells MAR1 and MAR2 (red circles), and 16 V.E.S. locations.

Groundwater with a salinity ranging from 0 to 3 g/l has been defined as freshwater, based on a study of the relationships between salinity and natural vegetation species richness [7]. Above the salinity threshold of 3 g/l, the species richness decreases dramatically. The brackish water has been defined as water with a salinity ranging from 3 to 15g/l and water with salinity above 15 g/l is considered saline.

The salinity data show that the size of the freshwater lenses changes throughout the year (Figure 13). Below the coastal pine forest and close to sea, the aquifer is salty and there is a 0.5 m thick fresh-brackish water layer at the top of the aquifer, which disappears in summer. Salinity data indicate a counter-intuitive trend underneath the irrigated agricultural land at 2 km from the sea: in summer the salinity is lower than in the winter. This may be explained by the high irrigation rates during the growing season or by infiltration from the irrigation ditch. Inside the historical *Classe* pine forest, at 4 km from the coast, the groundwater is mostly brackish (10 g/l) with a 1 m thick freshwater layer at the top of the aquifer that, in the coastal pinewood, vanishes rapidly. Fresh groundwater lenses within the quarry belt at 7 km from sea have a larger thickness (2 m) but at the bottom of the aquifer saltwater is also present. At a distance of 10 km from the coast, in agricultural land, there is fresh water throughout the whole thickness of the aquifer (Figure 13).

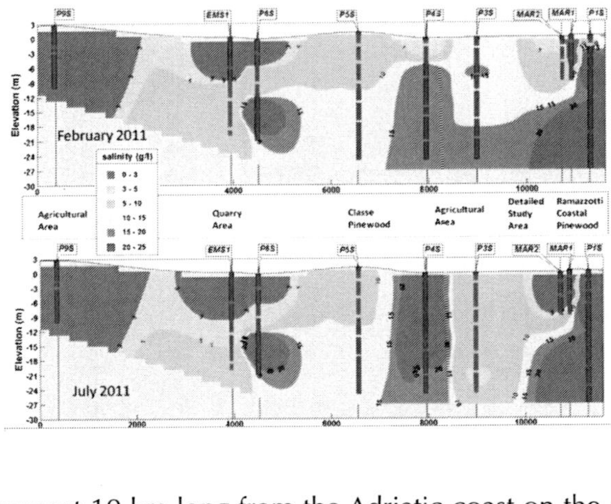

Figure 13. Transect 10 km long from the Adriatic coast on the right towards the mountains on the left. Warm colors represent brackish-saline water whereas cold colors are for freshwater.

On the basis of the cations and anions analyses, alkalinity, and salinity, a groundwater type classification used by Stuyfzand [45] was obtained [11]. Seawater is typically of the S4NaClo type meaning a saline water with High Alkalinity, Na being the major cation, Cl the major anion and a zero BEX indicating a stable water without cation exchange taking place (for data see [11]).

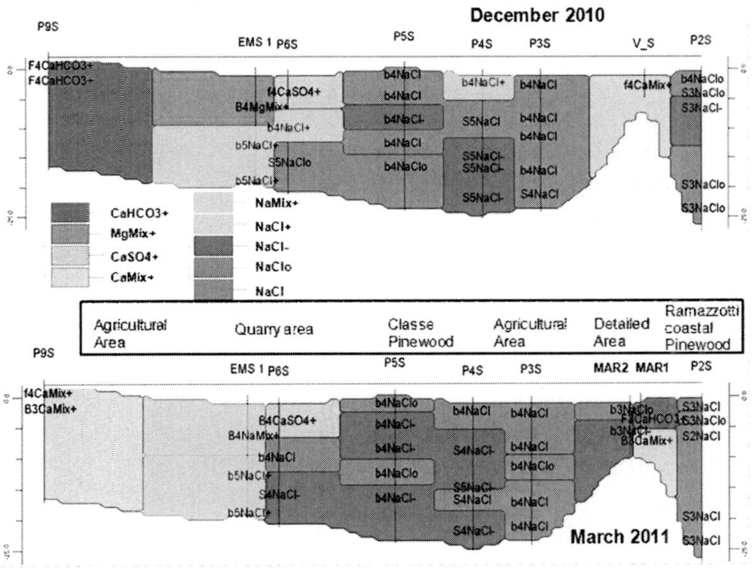

Figure 14. km-long transect from the Adriatic coast (on the right) towards the mountains (on the left). The different water types are indicated with different colors. Warm colors represent brackish-saline Na-Cl composition whereas cold colors are for typical freshwater composition.

Apennine river water is typically of the type (g-F)3$CaHCO_3$ and (f-F)3CaMix, or in other words: fresh water, medium alkalinity and rich in Calcium and varying anions. Po river water coming through the CER is of the type g2$CaHCO_3$ and slightly less alkaline than Apennine river water [11]. The groundwater water type is diagnostic to evaluate if the river-irrigation water has replaced the saline groundwater. Figure 14 shows that the sodium-chloride water type is dominant in the whole section. It is important to note that there are differences between the winter and spring situations in the area studied in detail. This was caused by the addition of new wells in the second monitoring campaign (MAR1 and MAR2 were monitored in April 2011 only). Excluding this effect, two water type changes were observed in the western part of the transect: from calcium-carbonate in December 2010 to calcium-mix in April 2011 in well P9S and from magnesium to sodium in well EMS. The April water type for the area near the irrigation ditch shows, in MAR1 well, a calcium carbonate water type, which is the same chemical composition of river water. Otherwise, in well MAR2 despite its low salinity (Figure 13), the dominant water type is sodium-chloride (Figure 14).

The BEX index for dolomite-rich aquifers as reported by Stuyfzand [45] is used to study whether the groundwater is becoming more or less saline. The BEX is indicated in Figure 15 with a +, - or 0 character after the water type code. The groundwater samples at the top of the aquifer and in winter show a positive BEX index, which indicates freshening (Figure 14).

Rainfall from January to September 2011 recorded in the five pluviometers along the section has a fairly homogeneous value around 400 mm. Total irrigation of a tomato field in the area studied in detail was 600 mm, applied with a rate of 120 mm per month (April-August). The infiltration rate of water into the aquifer, measured from drain gauges located in the tomato field, was 8 mm/m^2 in May, 2 mm/m^2 in June and 0 in the other months (Table 1).

Table 1. Precipitations, irrigation and infiltration measured in tomato parcel from January to September 2011

2011 (unit mm)	**Months**	January	February	March	April	May	June	July	August	September
	Rainfall	33	38	96	15	31	40	43	0	27
	Irrgation	0	0	0	150	120	150	180	90	0
	Infiltration	n.a.	n.a.	n.a.	8	2	0	0	0	0

Water table measurements in April 2011 show that the groundwater is at 0.2 m below sea level in the*Ramazzotti* pinewood and, moving landward, the depth linearly increases up to 1.2 m below sea level (Figure 15). The water level decreases rapidly everywhere during the summer, except for the area surrounding the irrigation channel (Figure 15) where an inverse gradient, with respect to the normal situation, is present during the whole irrigation time (from May to August 2011). As in winter the water level in the irrigation ditch is equal to the groundwater level, during the growing season the water level in the ditch is higher than the groundwater level, creating a hydraulic gradient in the phreatic aquifer that is directed to sea. At the end of the irrigation season (September for the year 2011), the inputs of freshwater end and the water level in the ditch returns to match the groundwater level (Figure 15).

The February 2011 survey results suggest that below the eastern parcel, where the aquifer is phreatic, a 4-m-thick freshwater lens is present before the irrigation season; freshwater is also present west of the ditch but its thickness is only 1,5 m. During both field campaigns in February and July 2011, VES profiles and the depth of the water table and the salinity of the groundwater at the water table were measured by drilling auger holes. The measurements made in February (Figure 16) show that the water level in the ditch is the same of the groundwater and that the whole system has a small gradient towards the pumping station. The freshwater lens vanishes below the *Ramazzotti* pinewood (E8 location) where the the water is salty up to the surface (Figure 16).

During the irrigation period (July 2011), the ditch is full of freshwater. The irrigation of tomatoes in the eastern parcel causes the whole soil profile to be moist. West of the ditch, where the Alfa Alfa was grown, the water table depth increases and also the bottom of the dry soil line becomes deeper (Figure 16). Comparing the VES profilqes made in February 2011 with those made in July 2011 confirms the existence of a well-developed freshwater lens in the eastern part that grows in size during the irrigation season (Figure 16). The largest growth in thickness of the freshwater lens occurs under the ditch and in the western part where the semi-confined aquifer is present.

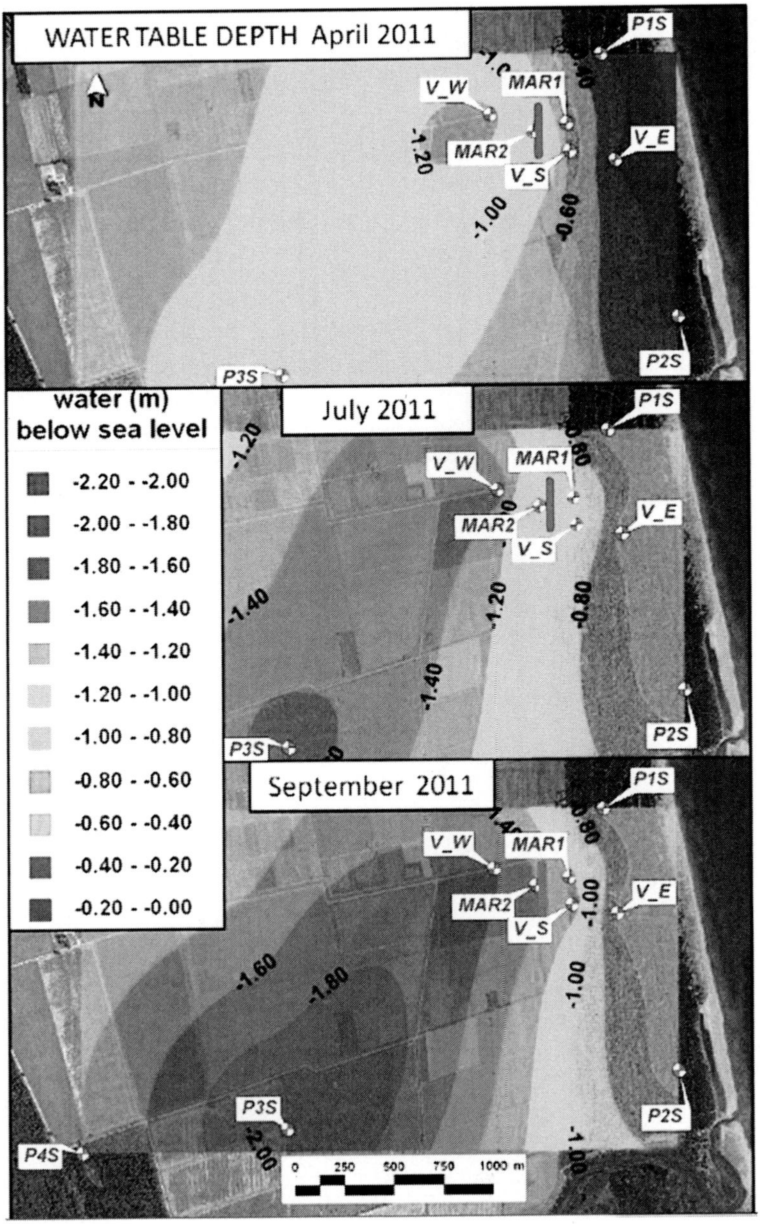

Figure 15. Study area water table map for April, July, and September showing also the irrigation ditch (red line).

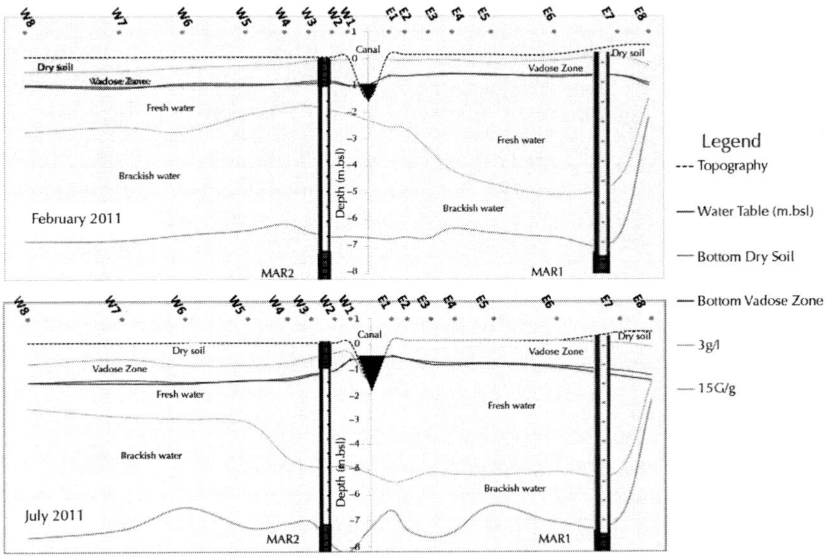

Figure 16. Seasonal comparison of VES results for the agricultural land close to the freshwater storage ditch.

DISCUSSION

As recently reported by IPCC [22] a reduction of freshwater availability and an increase of drought periods length is likely to occur in the Ravenna area during the next 100 years. An increase in temperature with a simultaneous increase of evaporation will cause soil salinization and loss of production. Near Ravenna more than 68 square kilometers are at risk for soil salinization caused by salt water intrusion in the coastal aquifer and along main rivers and channel. A stable isotope analysis shows that the ground and surface water near the city of Ravenna is a mix of sea, rain, and river water. The sea-water is, in part, inherited from the Holocene deepest marine ingression [11]. The aquifer has the shape of a sandy wedge enclosed in clay and this inhibits an efficient flushing of the saline water that is in part left at the bottom of the aquifer from the optimum climate period. Deeper aquifers in the area receive underground recharge from water with a mountain or rainfall sources. Where present, freshwater is located in the recent and paleo-dune belts, because of their high infiltration potential [3],[22],[47]. Where channels and rivers flow on a sandy top layer, there is infiltration of freshwater into the aquifer as along the *Bevano* and *Fiumi Uniti* rivers as well as in the western part of the San Vitale pinewood [11]. There is a small area along the *Bevano* River, where the freshwater lens reaches the bottom of the aquifer but in all other cases the freshwater lenses are floating on top of brackish groundwater. These freshwater lenses are thin (1m maximum), ephemeral (present only during winter) and strongly linked to the climate conditions [3], [22], [48], [49], [50].

Aquifer salinization in our study area is due to the combination of several factors such as low topography, natural and anthropogenic subsidence, mechanical drainage, destruction of the coastal dune belt, and scarcity of freshwater infiltration [11][43][51][52]). As shown in Figure 8, in the whole aquifer, the groundwater head is below sea level in winter also. This situation causes a sea-water encroachment promoted also by the high discharge water volumes of the pumping stations. Based on number of activity hours, we roughly calculated that in the *Quinto* Basin the volume of drainage water was 8 times larger than that of the total rainfall in same basin for the period.

In this contest, with low freshwater availability and poor natural recharge, it is important to look for some technique that is able to counteract soil and groundwater salinization and at the same time encourages water management practices that prevent nitrate leaching, safeguard the economics of agriculture, and save the freshwater resources.

Our attention was focused on a small agricultural area at a distance of 500m from the shoreline, where despite all previous studies [22][51][52][53], the presence of a freshwater lens throughout the year, was found. Although this area is affected by drainage, low topography, and subsidence as the rest of the Ravenna province, the presence of a sandy top layer, the horticulture with irrigation, and a ditch used for freshwater storage, all promoted aquifer recharge. The chemical analysis and electrical surveys showed that the size of the freshwater lens is larger during the summer period (Figure 13,Figure 14 and Figure 16).

In many other countries worldwide, managed aquifer recharge (MAR) is used to store fresh river or rain water in the underground of riverbanks or dunes for later use as drinking water [54].

Managed Aquifer Recharge or MAR is also used for re-pressurization of aquifers subject to falling water levels, declining yields, saline intrusion or land subsidence. Examples of these from around the world include different technologies such as aquifer storage and recovery (ASR), aquifer storage transfer and recovery (ASTR), bank and dune filtration, infiltration ponds, percolation tanks, rainwater harvesting, and soil-aquifer treatment (SAT) [55]-[61]. All above mentioned technologies have a cost that sometimes may be not sustainable by the local economy. In countries such as India, China, and Pakistan artificial recharge from ditches is affordable for the cultivation of rice, but this is usually done far from the coast ([61][62][63][64]). As reported by Gale 2000 ([61]), in many countries, the seepage from ditches or channels is considered a loss of water and they are studying systems to solve this problem. Irrigation ditches, on the other hand, may also play an ecological role: some authors report that ditches support the biodiversity and associated ecosystem services ([65]) and that they promote retention and removal of agrochemicals, decrease soil erosion, and allow the early stage of development for fish, amphibians, and reptiles [66]. Another aspect to consider is the low cost for their construction and management (removal of excess vegetation to restore efficient drainage) [66]. In particular for our area a dense network of ditches is already present and their building costs could be very low or absent.

Irrigation practices tend to become more and more efficient by letting go less and less water into soil-evaporation. The driving forces for that are the scarcity of freshwater and economic incentives. One might say that the water left in the ditch to recharge the aquifer is water that is 'lost'. One of the main challenges in water resources and agricultural management in coastal Mediterranean zones will be to define priorities that may vary with the particular area. In one place it may be important to use as little water as possible for irrigation not leaving anything to waste or end up in the aquifer underneath the top soil. In other places instead, especially in the coastal zone, irrigation water may be left to seep from the ditches into the surface aquifer to create freshwater lenses that sustain the agricultural land as well as the natural areas and at the same time counteract seawater intrusion.

The high salinity of the coastal aquifer near Ravenna is in part inherited from its geological history. Therefore it is unlikely that the aquifer will be ever used as a source for drinking water. By infiltrating (excess) river water on purpose through ditches as the one studied in this paper, however, freshwater lenses can be created that may prevent soil salinization and serve agricultural crops as well as natural areas. Our studied ditch has the same dimension as the drainage channels in the area. The only difference is that it is a closed basin used for storage irrigation water during the summer. By using the existing hydraulic infrastructure for managed aquifer recharge in ditches, more freshwater would flow into the aquifer instead of being drained away by the pumping machines. This system of aquifer artificial recharge would be viable in similar coastal settings throughout the world.

CONCLUSIONS

Despite the fact that there are many technological, agricultural, economic, and water availability factors that are dependent on the local situation and farming practices (crop rotations, volume and timing of irrigation, tillage, economic forcing, etc.) and that the length of the study was of only one year, so that no long term trends can be captured by the analysis presented, some interesting observations have been made:

- The land use in the Ravenna coastal area is complex; natural areas (pine forests, lagoons and wetlands) and industrial activities are placed in a landscape dominated by agriculture.
- The Low topography, the high rates of subsidence, the salty phreatic aquifer require land drainage to allow for any activity. The result is that drainage allows for agriculture but it decreases the recharge of the aquifer, removes freshwater from the top of the aquifer and increases up-welling of saline water from below.
- The surface hydrology is controlled by a high density network of channels used mainly for drainage but locally for irrigation.
- Due to different geological events, different kinds of sediments were deposited in the area; therefore special soils were developed and crops distribution as well as irrigation water requirements depends strictly from the soil characteristics.
- Irrigation is essential for horticultural crops, which are mainly located on sandy soil where the connection with the aquifer below is guaranteed.

- The water sprayed on the land for irrigation does not infiltrate into the aquifer and this is a good practice to avoid nitrate leaching and contamination of the groundwater. In the area studied in detail, the only irrigation water added to the aquifer is from seepage from the storage ditch.
- The case study shows that the presence of a ditch, full of freshwater for 5 months during the year, generates a permanent freshwater bubble 4 m in thickness, at 500 m distance from the sea. This shows that combined irrigation and drainage management, considering the dense network of existing channels and different types of soil, can be of benefit for the agriculture for soil protection, storage of freshwater resources, and to counteract sea-water intrusion.

ACKNOWLEDGEMENT

We thank the farmer of the "Cooperativa Braccianti di Campiano" for their helpfulness and kindness; we thank also University of Bologna, the Autorità dei Bacini Romagnoli and the Municipality of Ravenna for funding the several projects that form the basis of this chapter. Special thanks to Murugan Ramasamy and Sathish Sadhasivam for their fundamental support in field work.

REFERENCES

1. G. Toth, L. Montanarella, E. Rusco, 2008Updated Map Of Salt Affected Soils In The European Union. Threats To Soil Quality In Europe, Office For Official Publications Of The European Communities, Luxembourg (2008), 6174
2. P. G. B. Louw, S. De Eeman, B. Siemon, B. R. Voortman, J. Gunnink, E. S. Baaren, G. H. P. Van And, Essink. Oude, 2011Shallow Rainwater Lenses In Deltaic Areas With Saline Seepage, Hydrol. Earth Syst. Sci. Discuss., 876577707
3. M. Antonellini, P. Mollema, B. Giambastiani, E. Banzola, K. Bishop, L. Caruso, A. Minchio, L. Pellegrini, M. Sabia, E. Ulazzi, Gabbianelli. G. And, 2008Salt Water Intrusion In The Coastal Aquifer Of The Southern Po-Plain, Italy. Hydrogeology Journal 1615411556DoiS10040-008-0319-9.

4. E. Custodio, 2010Coastal Aquifers Of Europe: An Overview. Hydrogeology Journal, 181269280DOI:S10040-009-0496-1
5. FAO2002Agricultural Drainage Water Management In Arid And Semi-Arid Areas Ftp://Ftp.Fao.Org/Docrep/Fao/005/Y4263e/Y4263e11.Pdf
6. E. V. Maas, G. J. Hoffman, 1977Crop Salt Tolerance, Current Assessment J. Irrig. Drain. Div., 103, 115134
7. M. Antonellini, Mollema. P. And, 2010Impact Of Groundwater Salinity On Vegetation Species Richness In The Coastal Pine Forests And Wetlands Of Ravenna, Italy. Ecological Engineering. 236 (9), 1201-1211, Doi:10.1016/J.Ecoleng.2009.12.007.
8. I. Hutchinson, 1991Salinity Tolerances Of Plants Of Estuarine Wetlands And Associated Uplands. Washington State Department Of Ecology Report, Washington, USA.
9. IPWEA,2001The Local Government Salinity Management Handbook. Institute Of Public Works Engineering Australia (IPWEA), Australia.
10. T. Lombardi, B. Lupi, 2006Effect Of Salinity On The Germination And Growth Of Hordeum Secalinum Schreber (Poaceae) In Relation To The Seeds After Ripening Time. Atti Della Società Toscana Di Scienze Naturali, Memorie, Serie B 1133742
11. P. Mollema, A. Antonellini, E. Dinelli, G. Gabbianelli, N. Greggio, P. J. Stuyfzand, 2012Identification Of Fresh And Salt Water Intrusion Types In The Upper Ravenna Coastal Aquifer (Italy) Via Hydrochemical Systems Analysis. Submitted to Applied Geochemistry.
12. Istat2000Censimento Generale Dell'agricoltura http://www.census.istat.it
13. Provincia Di Ravenna2006Piano Territoriale Di Coordinamento Provinciale
14. V. Marconi, M. Antonellini, E. Balugani, E. Dinelli, 2011Hydrogeochemical Characterization Of Small Coastal Wetlands And Forests In The Southern Po Plain (Northern Italy). Ecohydrol. 4
15. P. Mollema, M. Antonellini, G. Gabbianelli, M. Laghi, V. Marconi, A. Minchio, 2012Climate And Water Budget Change Of A

Mediterranean Coastal Watershed, Ravenna, Italy. Environmental Earth Sciences. 65257276Doi:S12665-011-1088-7..

16. Fao.1990Guidelines For Soil Profile Description. Third Edition (Revised). Soil Resources, Management And Conservation Service, Land And Water Development Division, Fao, Rome.
17. Soil Survey Staff.1996Keys To Soil Taxonomy. Seventh Edition. United States Department Of Agriculture, Washington D.C.
18. D. A. Rose, F. Konukcu, J. W. Gowing, 2005Effect Of Water table Depth On Evaporation And Salt Accumulation From Saline Groundwater. Aust. J. Soil Res. 43565573
19. Andrej. T. Il'ichev, Tsypkin. And, G. George, Pritchard. And, And. David, Chris. N. Richardson, 2008Instability Of The Salinity Profile During The Evaporation Of Saline Groundwater. Journal Of Fluid Mechanics, 614. 871040022-1120
20. N. Filippi, 1994Carta Pedologica (Soil Map) 1: 250.000. Stampata Dalla Regione Emilia Romagna, Italy
21. S. Corticelli, 2003Carta Del Uso Del Suolo (Land Use Map). Copertura Vettoriale. Stampata Dalla Regione Emilia Romagna, Italy
22. B. M. S. Giambastiani, M. Antonellini, Essink. G. H. P. Oude, R. J. Stuurman, 2007Salt Water Intrusion In The Unconfined Coastal Aquifer Of Ravenna (Italy): A Numerical Model. Journal Of Hydrology 34094104
23. B. M. S. Giambastiani, P. N. Mollema, M. Antonellini, 2008Groundwater Management In The Northern Adriatic Coast (Ravenna, Italy): New Strategies To Protect The Coastal Aquifer From Saltwater Intrusion. In: Groundwater: Modelling, Management And Contamination. Authors: König L.F. And Weiss J.L.; Nova Publishers
24. Consorzio Di Bonifica Della Romagna2002Il Comprensorio E Le Infrastrutture http://www.bonificaromagna.it/Cartografia.ìt/html
25. Regione Emilia Romagna, Agip,2009Riserve Idriche Sotterranee Della Regione Emilia Romagna, Technical Report. Regione Emilia Romagna, Firenze
26. F. Giorgi, P. Lionello, 2008Climate Change Projections For The Mediterranean Region. Global Planet Change 6390104

27. G. Antolini, V. Marletto, F. Tomei, V. Pavan, R. Tomozeiu, 2008Atlante Idroclimatico Dell'emilia-Romagna 19612008
28. A. Magnan, B. Garnaud, R. Billé, F. Gemenne, Hallegatte. S. Iddri, Cired-Météo, 2009The Future Of The Mediterrenean. From Impacts Of Climate Change To Adaptation Issues. Institut Du Développement Durable Et Des Relations Internationals, Paris, France, 43 Pp.
29. A. Amorosi, M. L. Colalongo, G. Pasini, D. Preti, 1999Sedimentary Response To Late Quaternary Sea-Level Changes In The Romagna Coastal Plain (Northern Italy). Sedimentology, 4699121
30. R. Catalano, E. Di Stefano, S. Infuso, A. Sulli, P. R. Vail, F. P. Vitale, 1995Sequence And Systems Tracts Calibrations On High Resolution Bio-Chronostratigraphic Scheme: The Central Mediterranean Plio-Pleistocene Record. In: Mesozoic-Cenozoic Sequence Stratigraphy Of Western European Basins (Ed. By P.C. De Graciansky Et Al.). Spec. Publ. Soc. Econ. Paleont. Miner
31. A. Rizzini, 1974Holocene Sedimentary Cycle And Heavy Mineral Distribution, Romagna-Marche Coastal Plain, Italy: Sedimentary Geology, 111737
32. L. M. J. U. Van Straaten, 1970Holocene And Late Pleistocene Sedimentation In The Adriatic Sea. Geo/. Runsch. 60106131
33. Giambastiani B.M.S.2003Geomorfologia Del Litorale Ravennate: I Sistemi Dunosi. Unpublished Msc Thesis: University Of Bologna
34. F. Stecchi, A. Minchio, C. Del Grande, Gabbianelli. G. And, 2003Historical And Recent Evolution Of The Reno Rover Mouth And Adjacent Areas. In. Gabbianelli G. And Sangiorgi F. (Editors). 4Of Proceedings Of The Sixth International Conference On The Meditteranean Coastal Environment, Medcoast 03, 7-11 October 2003, Ravenna Italy, 5567
35. M. Bondesan, Gb. Castiglioni, C. Elmi, G. Gabbianelli, R. Marocco, Pa. Pirazzoli, A. Tomasin, 1995b Coastal Areas At Risk From Storm Surges And Sea Level Rise In North-Eastern Italy. J Coast Res 1113541379
36. A. Amorosi, M. L. Colalongo, F. Fiorini, F. Fusco, G. Pasini, S. C. Vaiani, Sarti, G.,2004Palaeogeographic And Paleoclimatic Evolution Of The Po Plain From 150-Ky Core Records. Global And Planetary Change, 405578

37. A. Correggiari, M. Roveri, F. Trincardi, 1996Late-Pleistocene And Holocene Evolution Of The North Adriatic Sea. Late-Glacial And Early Holocene Climatic And Environmental Changes In Italy: Il Quaternario: Italian Journal Of Quaternary Sciences 9697704
38. A. Amorosi, M. L. Colalongo, F. Fiorini, F. Fusco, G. Pasini, S. C. Vaiani, Sarti, G.,2004Palaeogeographic And Paleoclimatic Evolution Of The Po Plain From 150-Ky Core Records. Global And Planetary Change, 405578
39. L. Marchesini, A. Amorosi, U. Cibin, G. G. Zuffa, E. Spadafora, D. Preti, 2000Sand Composition And Sedimentary Evolution Of A Late Quaternary Depositional Sequence, Northwestern Adriatic Coast, Italy. Journal Of Sedimentary Research 70 (4), 829-838.
40. D. K. Bulter, 2005Near-Surface Geophysics, (Seg Investigations In Geophysics Series; N°13), Library Of Congress; Pag. 46
41. H. Burger, Robert, 1992Exploration Geophysics Of The Shallow Subsurface. Prentice Hall Pagg. 291295
42. P. Celico, 1998Prospezioni Idrogelogiche. Vol I, Liguori Editore, V° Ristampa 1998 Pag 243
43. D. Capo, 2011Phd Thesis "Caratterizzazione Della Falda Freatica Costiera Ravennate" In Press
44. M. H. Loke, 2001Electrical Imaging Surveys For Environmental And Engineering Studies- A Practical Guide To 2D And 3-D Surveys. Residinv Manual, Iris Instruments; 67 Pp.
45. P. J. Stuyfzand, 1989A New Hydrochemical Classification Of Watertypes. Iahs Publ. 1828998
46. IPCC Technical Paper Vi.2008Climate Change And Water-Intergovernmental Panel On Climate Change http://www.ipcc.ch/pdf/Technical-Papers/Climate-Change-Water-En.Pdf
47. V. Marconi, M. Antonellini, E. Balugani, E. Dinelli, 2011Hydrogeochemical Characterization Of Small Coastal Wetlands And Forests In The Southern Po Plain (Northern Italy). Ecohydrol. 4
48. B. Capaccioni, M. Didero, C. Paletta, L. Didero, 2005Saline Intrusion And Refreshening In A Multilayer Coastal Aquifer In The Catania Plain (Sicily, Southern Italy): Dynamics Of Degradation Processes According To The Hydrochemical Characteristics Of Groundwater. Journal of Hydrol 307116

49. Essink. G. H. P. Oude, 2001Improving Fresh Groundwater Supply-Problems And Solutions: Ocean And Coastal Management 44429449
50. Pp. Chot, Sc. Dekker, A. Poot, 2004The Dynamic Form Of Rainwater Lenses In Drained Fens: Journal Of Hydrology 2937484
51. V. Marconi, M. Antonellini, M. Laghi, A. Minchio, D. Savelli, 2008A Water Table Fluctuations Model In Sandy Soil Below A Coastal Pine Forest. Geophysical Researches Abstract 10: Egu201005761
52. Marconi V., PhD Thesis2010Hydrogeochemical Characterization Of Small Coastal Wetlands And Forests In The Southern Po Plain (Northern Italy).
53. V. Marconi, 2011Effetti Della Vegetazione E Del Drenaggio, Sull'intrusione Salina Nell'acquifero Freatico Costiero Della Zona Compresa Fra Foce Dei Fiumi Uniti E Foce Bevano (Ravenna). Dottorato Di Ricerca In Scienze Ambientali: Tutela E Gestione Delle Risorse Naturali, Ciclo XXIII. Università Di Bologna
54. P. J. Stuyfzand, 1999Patterns In Groundwater Chemistry Resulting From Groundwater Flow. Hydrogeology Journal 7 1527
55. Böhlke Jk2002Groundwater Recharge And Agricultural Contamination Hydrogeology Journal (2002) 10:153-179 DoiS10040-001-0183-3
56. J. Boonstra, M. N. Bhutta, 1996Groundwater Recharge In Irrigated Agriculture: The Theory And Practice Of Inverse Modeling Journal Of Hydrology 174 357 374
57. H. Bouwer, 199, 1997Role Of Groundwater Recharge And Water Reuse In Integrated Water Management. Arabian J. Sci. Eng., 22123131
58. H. Bouwer, 1999Chapter 24: Artificial Recharge Of Groundwater: Systems, Design, And Management. Hydraulic Design Handbook, L. W. Mays, Ed., Mcgraw-Hill, New York, 24.124
59. H. Bouwer, P. Fox, P. Westerhoff, Drewes. J. E. And, 1999Integrating Water Management And Reuse: Causes For Concern? Water. Qual. Internat., 19991922
60. P. Dillon, 2005Future Management Of Aquifer Recharge Hydrogeol J. 13313316

61. I. Gale, 2000Strategies For Managed Aquifer Recharge (Mar) In Semi-Arid Areas International Association Of Hydrogeologists Commission On Management Of Aquifer Recharge Iah- Mar International Hydrological Programme (Ihp) Unesco Division Of Water Sciences www.iah.org/recharge

62. K. S. Meijer, 2000Impacts Of Concrete Lining Of Irrigation Canals, Uda Walawe, Sri Lanka. M.Sc. Thesis, University Of Twente, The Netherlands.

63. K. Meijer, E. Boelee, D. Augustijn, I. Van Der Molen, 2006Impacts Of Concrete Lining Of Irrigation Canals On Availability Of Water For Domestic Use In Southern Sri Lanka. Agric. Water Manage. 83243251

64. Liqiang Yao, Shaoyuan Feng, Xiaomin Mao, Zailin Huo, Shaozhong Kang, D.A. Barry.2012Coupled Effects Of Canal Lining And Multi-Layered Soil Structure On Canal Seepage And Soil Water Dynamics Journal Of Hydrology 430-431, 91-102

65. R. B. Bradbury, W. B. Kirby, 2006Farmland Birds And Resource Protection In The Uk: Cross-Cutting Solutions For Multi-Functional Farming? Biological Conservation, 129, 530542

66. I. Herzon, J. Helenius, 2008Agricultural Drainage Ditches, Their Biological Importance And Functioning Biological Conservation 14111711183

Chapter 4

Impact Assessment and Multicriteria Decision Analysis of Alternative Managed Aquifer Recharge Strategies Based on Treated Wastewater in Northern Gaza

Mohammad Azizur Rahman [1,2], Bernd Rusteberg [2], Mohammad Salah Uddin [2,3], Muath Abu Saada [4,5], Ayman Rabi [4] and Martin Sauter [2]

[1]Institute of Fluid Mechanics and Environmental Physics in Civil Engineering, Leibniz University Hannover, Hannover D-30167, Germany

[2]Department of Applied Geology, Geoscience Center, Georg-August Universität Göttingen, Goldschmidtstr. 3, Göttingen D-37077, Germany;

[3]Irrigation Management Division, Institute of Water Modelling, New DOHS, Mohakhali Dhaka-1206, Bangladesh;

[4]Palestinian Hydrology Group, Emil Habibi Street-AlMasyoun, Ramallah, Palestine;

[5]Al Farez Center-South of the PA Compound, P.O Box 4288, Albireh, Ramallah, Palestine;

ABSTRACT

For better planning of a managed aquifer recharge (MAR) project, the most promising strategies should analyze the environmental impact, socio-economic efficiency, and their contribution to the existing or future water resource conditions in the region. The challenge of such studies is to combine and quantify a wide range of criteria from the environment and society. This necessity leads to an integrated concept and analysis. This paper outlines an integrated approach considering environmental, health, social and economic aspects to support in the decision-making process to implement a managed aquifer recharge project as a potential response to water resource problems. In order to demonstrate the approach in detail, this paper analysed several water resources management strategies based on MAR implementation, by using treated wastewater in the Northern Gaza Strip and the potential impacts of the strategies on groundwater resources, agriculture, environment, health, economy and society. Based on the Palestinian water policy (Year 2005–2025) on wastewater reuse, three MAR strategies were developed in close cooperation with the local decision makers. The strategies were compared with a base line strategy referred to as the so-called "Do Nothing Approach". The results of the study show that MAR project implementation with treated wastewater at a maximum rate, considered together with sustainable development of groundwater, is the best and most robust strategy amongst those analyzed. The analysis shows the defined MAR strategies contribute to water resources development and environmental protection or improvement including an existing eutrophic lake. The integrated approach used in this paper may be applicable not only to MAR project implementation but also to other water resources and environmental development projects.

INTRODUCTION

Nowadays, managed aquifer recharge (MAR) is considered as an integral part of integrated water resources management (IWRM). Like the IWRM concept, the interaction of MAR with other sectors of the water resources system, society, and natural processes is inherently strong [1]. Several researchers e.g., [2, 3] mentioned that like other IWRM projects, the most promising MAR strategy should study the environmental impact, socio-economic efficiency, and their contribution to the existing or future water resources problem in the region [3]. Proper investigation and planning of MAR projects is important for successful application and can lead to significant risk reduction (e.g., environmental, health) and overall project cost reduction by potentially reducing uncertainties during project implementation. Again, proper planning requires impartiality and transparency in the evaluation of MAR options, considering explicit assessment of feasibility and cost-effectiveness [4]. Up until now, very few research studies have performed an extensive integrated study that consider the potential impacts on the environment, health, economy and society due to MAR project implementation and which select the best project option after intensive impact assessment [5].

The Gaza Strip, located on the eastern coast of the Mediterranean Sea, is a region facing severe water resources problems [6]. Due to the hot and dry climate, little surface water is available. Water supply relies mostly on groundwater resources located in the Northern Coastal Aquifer of Gaza [7]. The Beit Lahia Wastewater Treatment Plant (BLWWTP), located at Northern Gaza Strip, has been dysfunctional for some time now and is creating severe environmental, socio-economic and agricultural impacts for the public health and the environment [8, 9]. A detailed description of the water resources problem at the North Gaza strip is given in Section 4.1. A three-phase 20-year project involving the construction of a new WWTP, called the North Gaza Wastewater Treatment Plant (NGWWTP), is planned to be located further to the south near the Israeli border (see Figure 1) [10]. The new wastewater treatment plant will involve MAR of effluents [11]. The Palestinian Water Authority (PWA), along with international support, decided to use practical, already established MAR technologies such as infiltration ponds with Soil-Aquifer Treatment (SAT) to replenish the coastal aquifer in order to meet the continually rising demand of water for domestic, industrial, and agricultural use in this water-parched region [12,13,14]. Decision support is required to identify the best MAR project option to implement in the study area.

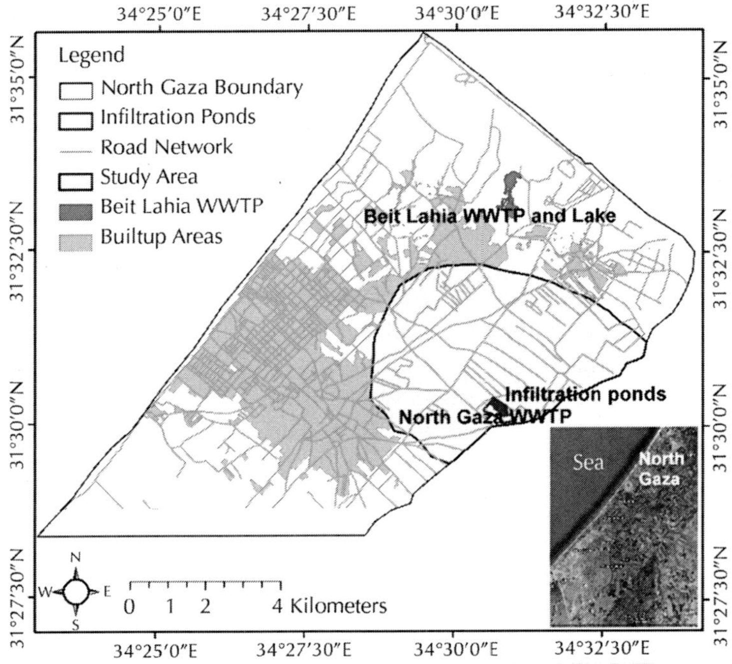

Figure 1. Study area map showing the wastewater treatment plants. Data source [9]. Inset picture from Google Earth.

In order to support the decision makers to plan the MAR project, this paper focuses on the impact assessment for several strategies for the implementation and operation of MAR in the Northern Gaza Strip. The strategies were quantitatively analyzed based on their potential impacts on agriculture, environment, health, society, and the economy. Finally, all strategies were compared to each other and ranked according to their ability to promote water resources development at the Northern Gaza Strip. In addition, this paper also describes the optimal MAR strategy of the candidates considered to sustain water resources and groundwater-dependent environment of Northern Gaza.

STUDY AREA

With an area of 365 km^2 and a population of roughly 1.6 million [9], the Gaza Strip is located on the southwestern part of historical Palestine at the Mediterranean Coast on the edge of the Sinai Peninsula. Precipitation varies between 200 and 400 mm/year, with an average of ca. 300 mm/year [6,15], and temperatures are generally high, ranging between 29 and 9 °C throughout most of the year [16], while 97% of water used in Northern Gaza comes from the Northern Coastal Aquifer [7]. In this study, a part of North Gaza was selected for analysis and comparison of MAR strategies (Figure 1), which is referred to in this paper as the "study area". The study area was delineated based on the boundary selection process using a groundwater flow and transport model. This model simulates the spreading of infiltration water at the new infiltration ponds, which commenced at the beginning of 2008 and will continue until 2040.

Geology and Hydrogeology of the Study Area

According to [17], the Gaza strip is underlain by a series of geological formations from the Mesozoic to the Quaternary. The two main formations are called Tertiary formation and Quaternary formation. The Tertiary formation, a 1200 m thick layer, is composed mainly of Saqiya formation and it consists of clay, marl and shale [14, 18, 19]. The 160 m thick Quaternary deposits covers the Pliocene Saqiya formation. The overlying Pleistocene deposits "Lower Quaternary" consists of (1) Marine Kurkur Formation (10–100 m thick on the coast); (2) Continental Kurkur Formation (maximum thickness is about 100 m with often-calcareous cement, and Quaternary Deposits. The sand loess and gravel beds formation is considered the main formation of the Gaza strip [17]. A general geological cross section of the coastal plain can be found in a number of sources [17, 20, 21, 22] and therefore is not included in this paper.

The North Gaza aquifer is a part of the Coastal aquifer that extends north to south from Haifa to the Sinai Coast. The highly permeable shallow vadose zone is mostly sand and gravel [23]. Larger and more consistent clay layers at the coast and extending 2–5 km inland, divide the Coastal Aquifer into several confined permeable layers [23]. The hydraulic connection between groundwater in the different subaquifers and the sea is not well investigated [17]. Beyond this distance, to the east, the Kurkar Group comprises the unconfined aquifer [18,23]. The average thickness of the aquifer at the coast is 150–200 m [23], whereas at the eastern border with Israel, the average thickness varies between 40 and 50 m [18]. The low-permeability Saqiya Formation of tertiary age constitutes the base of the aquifer. The 1 km thick Saqiya Formation is composed of clay, shale and marl [18]. The transmissivity of the Gaza aquifer ranges between 700 and 5000 m^2/d, corresponding hydraulic conductivity ranges between 20 and 80 m/d. Specific yield and Specific storativity values are 0.1–0.3 and 1×10^{-4} per meter [19, 24].

Rainfall is the main recharge component for the shallow aquifer unit in the study area. Aish *et al.*, (2009) [20] estimated that the average annual recharge of the Gaza strip is 108 mm/year (39–40 Mm^3/year). Around 1016 agricultural wells pump ca. 50 Mm^3/year and 45 urban supply wells abstract approximately 42 Mm^3/year. Irrigation return flow is considered as 30 Mm^3/year [18]. In the Gaza strip, the groundwater abstraction from the drinking water wells constitute more than 50% of the net withdrawal [25]. In the northern part of Gaza, groundwater levels range from about 2 m above MSL at the eastern border with Israel to mean sea level along the shore [18]. A steep groundwater level gradient is seen at the southern part of the Gaza strip. The coastal aquifer possesses 5000 Mm^3 storage of groundwater of variable quality of which 30% is fresh [26,27]. In North Gaza, the GWL in the centre of the area is lower than the other parts of the area. So, in this part of the coastal aquifer, the main groundwater flow direction is towards the centre of North Gaza [28]. Besides the water quantity shortage, groundwater quality related problems, especially chloride and nitrate contamination, have been mentioned by several researchers e.g., [18,19,29]. The existing monitoring network in the Gaza strip observes groundwater level, and measures nitrate and chloride concentrations. The network is not suitable for monitoring sea water intrusion [18].

METHODOLOGY

An integrated approach was formulated in order to select the best strategy for MAR implementation. The approach is integrated in the sense that the study considered the impacts of possible MAR strategies on several sectors such as environment, health, economy and society. The sequential steps to select the best rank MAR strategy, a structured and sequential work flow was prepared, as shown in Figure 2. In general, the entire process involves three main steps to identify the best ranked MAR strategy: (a) water resources system analysis and strategy development (b) strategy ranking: criteria selection, impact assessment and criteria quantification, and (c) Multicriteria decision analysis (MCDA).

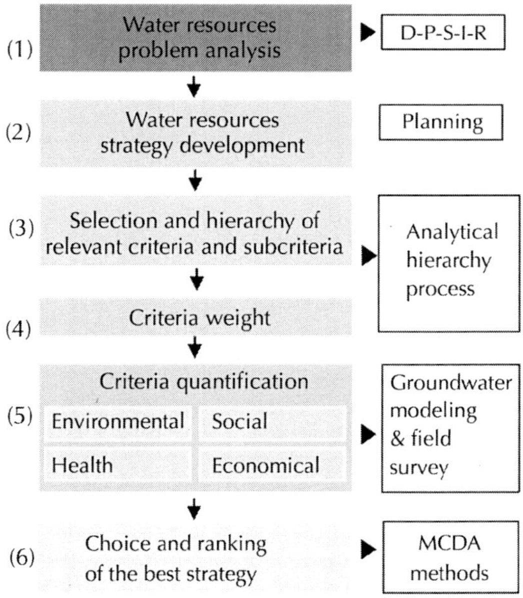

Figure 2. Overall methodology of the study.

The main objective of water resources system analysis (step-1) is to identify the main water resources drivers and pressures, and the potential responses to solve the impacts. Causal chain analysis using the Driver (D), Pressure (P), State (S), Impact (I) and Response (R), in short DPSIR, methodology [30,31] can be used at this step. Based on the pertaining water resources problem and the potential responses, water resources strategies are developed (step-2). The strategies should comply with the national water policy. In the third step of strategy ranking, relevant environmental, health, social and economic characteristics are selected. Each characteristic is defined as a criterion. The next step involves the decomposition of the ultimate goal into a hierarchy of several levels, following the principle of Analytical Hierarchy Process (AHP). The bottom level is the most specific criteria and the middle levels are more general criteria and can be called the "main criteria". The criteria in the lowest level are related to the main criteria in the middle levels. All levels combined is the goal of the study—the best strategy for MAR implementation, and is positioned at the top of the hierarchy. The next step in the strategy ranking procedure is assigning values of relative importance for each criterion at all levels, which is done by assigning a weight to each criterion. The criteria under each main criterion are compared amongst themselves and a weight is assigned to each one (step-4). The main criteria are also weighted in this way. The next step (step-5) is to quantify the relevant criteria, which is the main focus of the present study. A number of techniques, such as groundwater modelling, GIS and field surveys are available to quantify scores for the criteria. The quantification procedure depends on the type of criterion. After quantifying all criteria, an evaluation matrix is prepared at this step which is one of the principle components for ranking of alternatives. The final step (step-6), strategy comparison and ranking analysis, encompasses two multi-criteria analysis techniques: Weighted Linear Combination (WLC) and PROMETHEE II (Preference Ranking Organisation MeTHod for Enrichment Evaluations) method.

The role of AHP, mentioned earlier, was to construct the hierarchy and to estimate the relative weight by pairwise comparison, after getting the preference information from the researchers, decision makers and stakeholders. Additionally, the role of WLC and PROMETHEE is to rank the alternatives.

WATER RESOURCES PROBLEM ANALYSIS AND STRATEGY DEVELOPMENT

Water Resources Problem Analysis (Step-1)

With the aim to analyse the existing water resources problems of the study area, causal chain analysis using the DPSIR method was used. The DPSIR concept has been developed for describing interactions between society and the environment [31,32], starting from the assumption that there is an interaction between the two. The water resources problems of North Gaza were analyzed, decomposed, and structured in this method in order to find the potential response of the problem. In brief, the water resources system of North Gaza is affected by two main drivers: population and urbanization. These drivers cause certain pressures on groundwater exploitation, wastewater status, land-use change, salinization, *etc.* The causal chain analysis of surface water is negligible as there are no surface water resources in the area. The DPSIR analysis has identified four potential responses to the current water resources problem. Each response can be considered and studied independently as well as in combination. In this paper, we considered MAR as a potential response due to the following reasons: (1) the poverty level in Gaza is high and many cannot afford the costs of advanced water treatment or desalination (considered as innovative technology) [33]; (2) Treated wastewater reuse will complement the existing water resources and will improve the water supply for agriculture; (3) Use of reclaimed water for agriculture would make fresh groundwater available for domestic and industrial use. In this study, MAR is seen not only as a contribution for a solution to the water supply and groundwater quality issue, but also as a solution to the problematic effluent lake, located at Beit Lahia, as the use of the new infiltration pond would help to rehabilitate the old infiltration lake.

Water Resources Strategy Development (Step-2)

Based on the water resources problem analysis and considering the water resources management plans for the years 2005–2025 [2, 5, 9, 10], the following four MAR strategies were established in this study (Table 1).

The water management strategies based on MAR presented in Table 1 consider three phases in terms of wastewater resources development at the case study area. Strategy No.1 (Sc-1) represents the strategies if nothing has been changed with respect to the existing water resources structure and no further planning is being considered. Strategy No.2 (Sc-2) is linked to the first phase. This phase considers the diversion of the water from the BLWWTP to the newly constructed infiltration basin, which is located close to the foreseen position of the new North Gaza Wastewater Treatment Plant (NGWWTP) at the Israeli border. The diversion of water will be accomplished via a pressurized pipeline and the effluents will then infiltrate into the aquifer. Strategy No.3 (Sc-3) considers the strategies if the diverted water will be treated in the NGWWTP and then infiltrated into the aquifer. The effluent quality is higher than that of the water used for infiltration in Sc-2. In Phase 3, the NGWWTP is designed to increase the treatment capacity of around 24 Mm^3 per year in 2025. It indicates in Sc-3, the effluent water quality is better than that in Sc-2. Strategy No.4 (Sc-4) considers infiltration of this extra volume of treated water to the aquifer. Sc-2, Sc-3, and Sc-4 are considered as MAR management strategies.

Table 1. MAR management strategies towards the development of water resources at the Northern Gaza

Strategy No.	Plan for Water Resources Development	Scenario	MAR Volume (Mm^3) in Year 2040	Chloride/Nitrate Concentration in Recharge Water (mg/L)
Sc-1	Do Nothing	No MAR	0	557–887/20–107 *
Sc-2	Phase 1: Infiltration ponds and pipeline construction	Use the water from the BLWTTP	13	250/19–43

Sc-3	Phase 2: Construction of the NGWWTP	Infiltration of better quality water from the new treatment plant	13	250/7.5–17
Sc-4	Phase 3: Extension of the NGWWTP	Infiltration of better quality water and increase in infiltration volume from the new treatment plant.	23.7	250/7.5–17

CRITERIA SELECTION AND QUANTIFICATION PROCEDURE

Criteria Selection (Step-3)

A wide range of indicators are considered for the selection of criteria. The criteria were derived from the identified sectors of impact and emphasis was given to the availability of information to quantify the criteria. A total of 19 most representative decision criteria were selected in close cooperation with Palestinian researchers and authorities as well as further relevant stakeholders and were discussed with other international experts in related fields. Among the 19 criteria, six criteria represent environment considerations. They consider groundwater level, chloride and nitrate concentration averaged year 2005–2040 and also in year 2040 alone. Four health criteria consider chloride and nitrate concentration at the domestic wells average 2005–2040 and also in year 2040 alone. Seven social criteria consider people's acceptance, convenience, satisfaction with the water quality and quantity, employment and willingness to pay. Affordability to pay and net cost-benefit analysis were considered as economic criteria.

Figure 3 shows the four-level hierarchical structure of the categories and criteria. AHP was used at this step. The AHP, proposed by [34], is a multicriteria analysis technique that enables the explicit ranking of tangible and intangible factors against each other for the purpose of decision-making or conflict resolution. It combines qualitative and quantitative approaches [35].

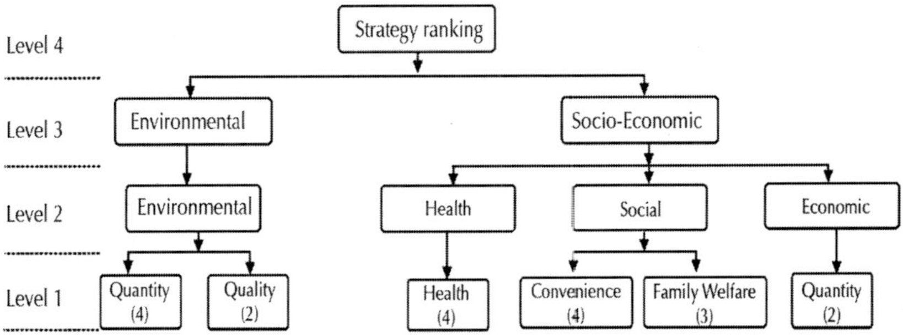

Figure 3. Criteria selection and hierarchy. Italic numbers indicate the number of criteria associated to each item at the fourth level.

Nineteen criteria were grouped into four main criteria groups such as environmental, health, social and economic. At the third level of the hierarchy, social, health, and economic criteria were grouped as "socio-economic" criteria. Socio-economic criteria and environmental criteria group combines the ranking of the strategies.

Criteria Weighting (Step-4)

The relevant importance of each criterion was defined in close cooperation with local scientists, decision makers and stakeholders. A participatory process was undertaken among the local stakeholders and experts. The participatory process includes scientific meetings, questionnaire surveys and workshops. Judgments of international experts were considered along with the weights from local experts and stakeholders. The pairwise comparison method, originally proposed by [34], was used to transfer the linguistic importance to numeric value and relative weights were estimated. The net cost and groundwater quantity were considered to be the most important criteria. All categories at level 2 and level 3 were considered as being equally important for MAR planning and management.

Criteria Quantification (Step-5)

The selected criteria were quantified using several state-of-art analysis techniques such as groundwater flow and transport models, field surveys, economic models, etc.

Quantification of Environmental Criteria (Criteria 1 to 6)

The selected environmental criteria refer to the groundwater quality and quantity status. These criteria were quantified by using groundwater-modelling techniques. A groundwater flow and transport model, developed in this case study using VISUAL MODFLOW (version 4.3, SWS, Vancouver, BC, Canada, 2009) and its integrated modules, was used to quantify the six environmental criteria in this study. The detailed description of the flow model set up and model parameters together with calibration plot can be found in [28]. The transport parameters such as longitudinal and vertical transverse dispersivity were initially assigned values of 4 m and 1 m, respectively (according to [36]). Bulk density of water was considered as 1000 kg/m3. For Sc-2 and Sc-3, the infiltration starts in 2008 with 9.7 Mm3 of treated water and with an increase of infiltration by 0.08 Mm3 per year until 2012 and afterwards the infiltration volume remains 13 Mm3 until 2040. For Sc-4, the infiltration starts in 2008 with 9.7 Mm3 of treated water and with an increase of infiltration by 0.08 Mm3 per year until 2040. During the analysis and quantification of all the strategies, the current water withdrawal for agriculture was assumed to be constant. Domestic water demand was assumed to increase (based on population growth), according to the estimated demand increase. The model was run until year 2040. Simulation results flow and transport modelling from years 2005–2040 were used to estimate Criteria 01, Criteria 03 and Criteria 05. Simulation results from flow and transport modelling at the end of year 2040 were used to quantify Criteria 02, Criteria 04 and Criteria 06.

Quantification of Health Criteria (Criteria 7–10)

The four health-related criteria refer to the water quality status at the domestic water supply wells. Average chloride and nitrate concentration were considered at the places where the domestic wells are situated (Criteria 07 and Criteria 08). Criteria 09 and Criteria 10 were quantified by considering the average concentration of chloride and nitrate in the waters of the study area aquifer. The developed groundwater flow and transport model was also used to quantify the health criteria for the analysis. The water quality in the domestic wells depends on the quality of infiltrated water, quality of native groundwater and the seasons (winter and summer). These three aspects were considered in the model.

Model Simulation for the Health Criteria Quantification for the Strategies

Chloride

Chloride was modelled as a conservative parameter and hence, no sorption or kinetic reaction was considered. The initial concentration, ranges between 40 and 2200 mg/L, of chloride was taken from the trend analysis in [37], considering the data from the years 1984–1998 [37,38]. The chloride concentration of the infiltrated water was considered to be the same as that in the wastewater lake at BLWWTP. The chloride concentration used in the model and during the entire modelling period was 559–857 mg/L for years 2004–2007 and 250 mg/L for years 2008–2040 in all strategies except Sc-1 [9]. For Sc-1, the base condition was maintained. The base condition considers the chloride concentration used in the simulation model from year 2000 to year 2003. The effect of chloride concentration changes as the volume of infiltrated water changes in different scenarios.

Nitrate

For nitrate simulation, equilibrium controlled linear isotherm was considered and no kinetic reaction was considered. Similar to chloride, the initial concentration, ranges between 5 and 370 mg/L, of nitrate was taken from the trend analysis from [37] and considered is the data from 1984 to 1998. The nitrate quality of the infiltrated water was calculated based on the quality of the infiltrated water, the infiltration process, and seasonal climatic conditions (after [37,38]) and where dilution and denitrification have been assumed to be the main processes for nitrate reduction in the model simulation. For Sc-1, a base condition was maintained throughout the entire simulation period. A base condition maintains the nitrate source, considering the same land use utilized in the simulation model 2000–2003. The nitrate concentration for Sc-2 used in the model and during the entire modelling period was 20–107 mg/L for years 2004–2007 and 19–43 mg/L for years 2008–2040. The nitrate concentration for Sc-3 and Sc-4 used in the model and during the entire modelling period was 20–107 mg/L, 19–43 mg/L and 7.5–17 mg/L for the period of 2004–2007, 2008–2011, and 2012–2040, respectively.

Quantification of Social Criteria (Criteria 11 and 17)

A questionnaire survey was performed by the Palestinian Hydrology Group to get the social aspect of the MAR strategies [33]. The questionnaire was prepared in such a way that it includes criteria that would measure the anticipated level of convenience, perceptions on willingness to use the recharged water for different purposes and the fees that the user would be willing to pay for the supply and the expected level of satisfaction from the quantity and quality of water supplied from each option. A total of 76 questionnaires were filled out by the locals in the area [33]. The number of questionnaire was decided based on statistical analysis and population residing at the study area.

Quantification of Economic Criteria (Criteria 18 and 19)

In the present study, two economic criteria were considered. Affordability to pay (criteria 18) was quantified using the surveyed data. Criterion 19 considers the net present cost and benefit of the four strategies implementation. For net present cost and benefit estimation, the following factors were considered (after [37]):

- The infiltration starts in 2008 with 9.7 Mm3 of treated water and with an increase of infiltration by 0.08 Mm3 per year according to the strategies.
- The estimated operation & maintenance (O & M) cost (water transfer, pumping of water, cleaning of infiltration basin etc.) for MAR is $0.14/m3.
- The cost of abstracting recharged water by wells is $0.11/m3.
- The cost of the land (80,000 m2) for the infiltration basin is $100,000 and was considered at the beginning of 2005, as the ponds were planned to be constructed in this year.
- The cost of construction of the nine infiltration ponds and water-pumping infrastructure is $4,000,000 and was considered in the estimation at the beginning of 2005.
- The opportunity cost will be represented mainly by the land that will be used to construct the infiltration basins. Since the area is an agricultural area and the net return per 1000 m2 (1 dunam) from various agricultural products (mainly vegetables and citrus) per year is $562, then the opportunity cost of the land (80,000 m2) is $44,960. The lake or the lagoon is planned to close down by year 2018. The area occupied by it is 100,000 m2. Considering the area will be used for agricultural production, it would produce an annual benefit $56,200 per year starting from the year 2018.
- The gains from improving water quality is calculated as the cost of desalinating brackish water of 30% of the private well if the MAR is not implemented (Sc-1). The cost of desalinating brackish water is considered as $0.36/m3.
- The cost of abstracting ground water by wells is $0.11/m3 due to groundwater lowering in Sc-1 after 2007.
- As a safety measure, $0.01/100 m3 of recharged water was considered as unforeseen cost due to implementation of MAR.
- The net benefit from the stored water was estimated considering the people's willingness to pay ($0.37/m3).
- The discount rate to calculate net present value was assumed to be 3% and assumed to be constant over all years of the project.

- No cost for wastewater treatment facilities was considered, as the local authority already considered this cost during the economic feasibility of the NGWWTP [14].

The cost estimation was done using an economic model based on a spreadsheet.

MCDA Analysis and Ranking of Options

After quantification of all the criteria, the normalized matrix was prepared for multicriteria analysis. The normalization was done using the following formula:

$$NV = \frac{\text{Max} - \text{Value}}{\text{Max} - \text{Min}} \quad (1)$$

Here, NV denotes normalized value, Max and Min indicate the maximum and minimum value among the values to be normalized, respectively. We use Equation 1 to normalize all criteria values.

Criteria Aggregation Methods: Weighted Linear Combination (WLC)

WLC combines the criteria and provides the ranking. WLC is the most simple and commonly used aggregation method in decision analysis [35].

$$S(xi) = \sum wj \cdot sj(xi) \quad (2)$$

where, w_j is a normalised weight; and $w_j = 1$; and $s_j(x_i)$ is the normalised criteria function.

After receiving the criteria weights and preparing the evaluation matrix, the role of WLC is to perform weighted summation for each group of criteria at all levels of the hierarchy until the strategy ranking achieves.

PROMETHEE

PROMETHEE, developed by [39], is a nonparametric outranking method for a finite set of alternatives. The method was later extended by [40,41]. PROMTHEE I gives partial ranking and PROMETHEE II provides a complete ranking of the strategies by using the net flow [42]. The details of the procedure can be found in many sources such as [39, 43, 44, 45].

RESULTS ANALYSIS

Environmental Criteria

The simulations show (see Figure 4a) that the maximum average GWL rise in the study area is 6 m by the year 2028 with respect to "Do nothing" (Sc-1). At the end of 2040, the GWLs are estimated to be –2.61 m, 0.81 m, and 3.57 m above sea level (ASL) for Sc-1, Sc-2 & Sc-3, and Sc-4, respectively. 3%–5% of the infiltrated water may flow to Israel each year under the simulation condition of Sc-2 and Sc-3, whereas this outflow was estimated to be 7%–15% per year for Sc-4. The inflow to the study area from the Israeli side will be reduced by 20%, for both Sc-2, Sc-3 and by 30% for Sc-4. Due to the infiltration of treated wastewater, the groundwater level below the infiltration basin would increase and would cause the fresh water flow to be reduced from the Israeli side.

Figure 4b shows the average chloride concentration in the study area for the four strategies. The model results show the average chloride concentrations at the end of 2040 are 522 mg/L, 426 mg/L, and 400 mg/L for Sc-1, Sc-2 & Sc-3, and Sc-4, respectively. Figure 4c shows the average nitrate (expressed as NO_3-N) concentration in the study area for the four strategies. The average nitrate concentrations at the end of 2040 are 82.27 mg/L, 67 mg/L, 59 mg/L, and 44 mg/L for Sc-1, Sc-2, Sc-3, and Sc-4, respectively. Implementation of Sc-4 will therefore provide storage in the aquifer with a maximum value of 23 Mm^3 per year after the full implementation of north Gaza wastewater treatment plant (NGWWTP), Phase 3 (year 2025).

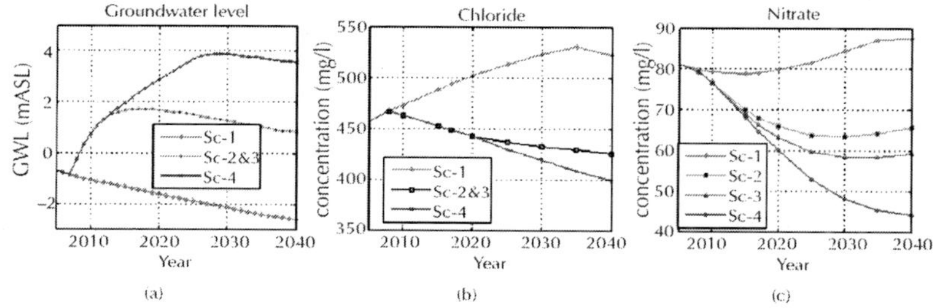

Figure 4. (**a**) Average groundwater level; (**b**) average chloride concentration and (**c**) average nitrate concentration in the study area during year 2005 to year 2040 for the four MAR strategies.

Health Criteria

A total of 10 domestic wells are located within the study area. Figure 5a shows the average chloride content of the 10 domestic wells for the four strategies until the year 2040. The average chloride concentrations at the end of 2040 are 555 mg/L, 528 mg/L and 407 mg/L for Sc-1, Sc-2 & 3, and Sc-4, respectively. In the case of Sc-1, the average chloride concentration in all domestic wells increases until the year 2040. In the case of Sc-2 & 3 and Sc-4, the average chloride concentration increases until the year 2035 and 2030, respectively, and then the chloride concentration decreases. Figure 5b shows the average nitrate content of the 10 domestic wells for the four strategies until the year 2040. Minimum nitrate concentration was observed in case of Sc-4. The average nitrate concentrations at the end of 2040 are 90 mg/L, 72 mg/L, 68 mg/L, and 49 mg/L for Sc-1, Sc-2, Sc-3, and Sc-4, respectively.

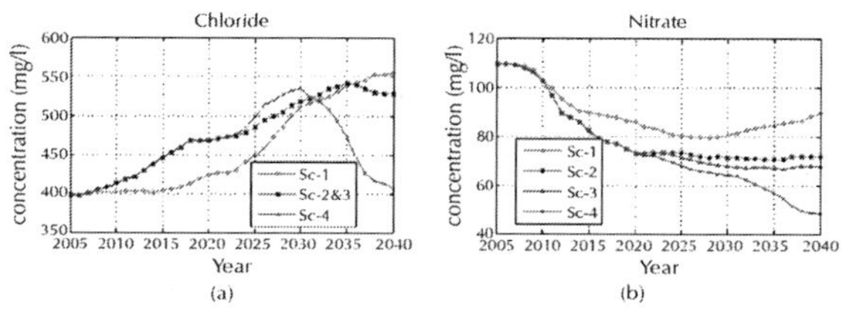

Figure 5. (**a**) Average chloride concentration; (**b**) Average nitrate concentration in the ten domestic wells for the entire simulation period (year 2005 to year 2040).

Social Criteria

The survey results indicate that 86% of the respondents agreed to reuse wastewater for agricultural purposes whereas 67% and 42% agreed to reuse wastewater for industrial and domestic purposes, respectively. Results also show that respondents are willing to pay very little for the infiltrated water regardless of use and claim to be able to afford very small fees. The inhabitants are willing to pay a maximum $0.37/m^3 to use wastewater for irrigation (Figure 6). The survey results indicate that the distribution of acceptance and satisfaction of the public is similar throughout the various MAR strategies. In terms of satisfaction with the water quality, perceptions range from being satisfied to fairly satisfied with Sc-3 and Sc-4 having the greatest level of satisfaction.

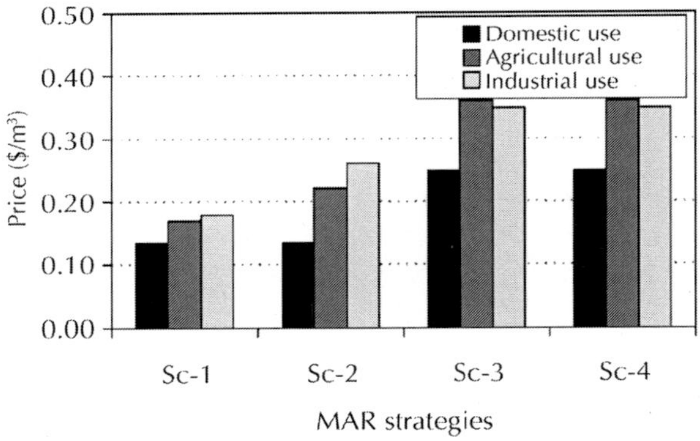

Figure 6. Willingness to pay of the respondents for the MAR strategies for different usage.

Economic Criteria

In the study area most of the people depend on agriculture, and many youths and women participate in agricultural activities. The agricultural activities in the study area depend on the groundwater irrigation. Hence, it is important to carefully review the water price (tariffs) for project feasibility. The survey results indicate that the respondents cannot afford to expend more money in order to use the benefit gained due to implementation of Sc-2, Sc-3 and Sc-4.

High investment cost is an important factor that makes a big difference between MAR strategies (Sc-2, Sc-3 and Sc-4) and the "Do nothing approach" (Sc-1). From the net benefit (cost-benefit) estimation (Figure 7), the implementation of a MAR strategy would be beneficial after year 2022 in case of Sc-4 and after year 2024 in case of Sc-2 and 3 (Figure 7). Sc-4 returns the most benefit due to its extended amount of infiltration volume even after year 2012. The net present values of the strategies (years 2005–2040) are \$10.2 M for Sc-2 and Sc-3 and \$28.4 M for Sc-4 whereas for Sc-1 the value is –\$32. 0 M (negative sign indicates net cost). That is, there is a \$60.4 M PV net benefit of switching from strategy Sc-1 to Sc-4 or a \$42.2 M PV net benefit of switching from Strategy Sc-1 to either Sc-2 or Sc-3.

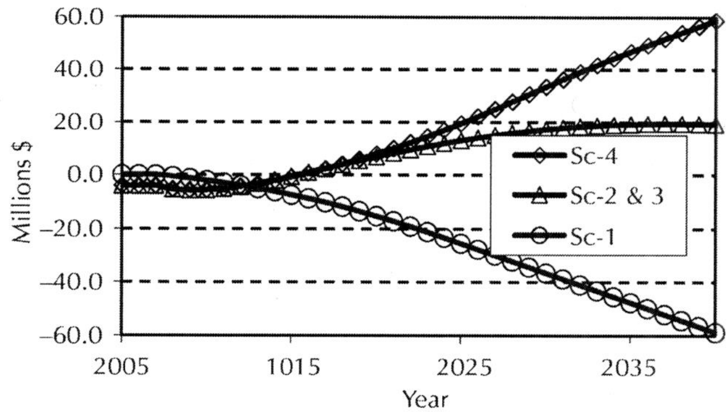

Figure 7. Net benefit analysis for the four MAR strategies.

STRATEGY COMPARISON AND RANKING

Figure 8 shows the performance of the four strategies according to the main criteria group (level-2). It is clear from the figure that Sc-4 performs the best in environmental, health and social criteria and Sc-1 performs the worst in these cases. Sc-2 performs better that Sc-3 according to the social and economic criteria but performs worse than Sc-3 for environmental and health criteria. People's affordability, convenience, and acceptance of wastewater seem important for the ranking. The final ranking was achieved after combining the main criteria groups (level-4) and the ranking is Sc-4 > Sc-3 > Sc-2 > Sc-1.

It was found that Sc-4 performs best for all the quantified detailed criteria with the following exceptions; average chloride concentrations in domestic wells over the study period, satisfaction with domestic water quality, willingness to pay and affordability to pay. These deviations are due to temporarily increased salinity of domestic wells in specific locations due to changed flow directions and variable salinity in the aquifer. This also influences criteria for satisfaction with domestic water quality for users of those domestic wells, and willingness to pay. Sc-4 also has the highest capital costs of all options (affecting affordability to pay), although the net benefits are greatest. For these specific criteria, only the "Do Nothing Case" (Sc-1) performs the best, although for other criteria it performs very poorly compared with other options, especially Sc-4.

PROMETHEE I partial ranking also confirms that Sc-4 performs better than the other strategies. No out-ranking relation does exist between Sc-2 and Sc-1; and Sc-1 and Sc-3. PROMETHEE II ranking is similar to that observed using WLC method.

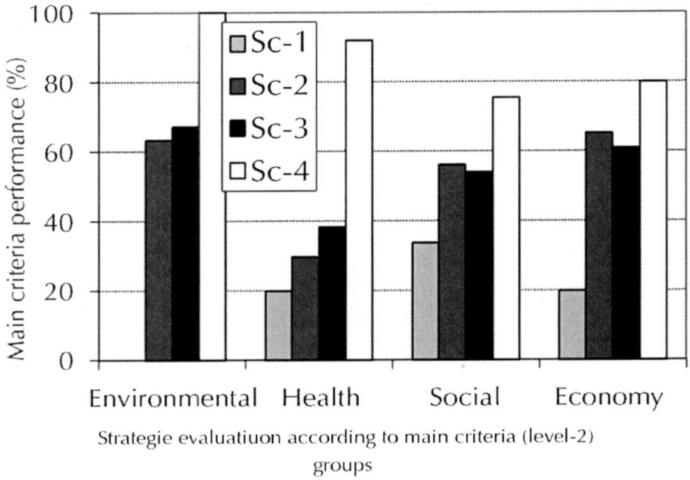

Figure 8. Ranking of the strategies according to main criteria group (level 2) using AHP-WLC combination.

DISCUSSION

Criteria Quantification

Environmental Criteria

The Sc-1 ("Do Nothing Approach") indicates continuous groundwater level mining over time, whereas Sc-4 indicates higher groundwater development than the other three strategies. Similarly, among the four strategies, Sc-4 shows better conditions in terms of inflow from the sea to North Gaza. Infiltration of excess treated wastewater even after 2012 might help Sc-4 to get more environmental benefit. In general, the problem of water flow from the sea will remain under control by the infiltration of all MAR strategies. It is clear from the results that Sc-1 ("Do Nothing Approach") will lead to deterioration of groundwater quality (*i.e.*, chloride and nitrate increase) with time. However, for other strategies, the groundwater quality will improve with time. The long-term effect of groundwater flow might also control the groundwater quality in the study area as the distribution of chloride and nitrate in North Gaza and the nearby Israel border is complex.

From the groundwater model simulation, we delineated a zone of ca. 200 m from the edge of the infiltration basins receiving the infiltrated water with a residence time of ca. six months. Regarding pathogenic bacteria, residence time of more than 6 months is recommended [46]. In the study area, no domestic wells exist within these 200 m.

Health Criteria

The impact of managed aquifer recharge projects on domestic wells is very sensitive to the population living in the area. The simulation result for Sc-4 shows a significant chloride concentration decrease in the study area at the end of the year 2040 in comparison to Sc-1, Sc-2 and Sc-3. By analysing the chloride concentrations in all domestic wells and comparing them with the "Do Nothing" strategy, observations show that the impact on chloride concentrations in all wells will be almost the same. Due to the groundwater flow direction of the infiltrated treated effluent, this would also impact the domestic wells. The increasing trend in the domestic well chloride concentration is due to the higher chloride concentration in the infiltrated water than the native groundwater and groundwater flow direction. In general, the nearby aquifer of the wells and the aquifer beneath the infiltration basin display higher chloride concentration. The infiltrated water would displace this water towards the domestic wells and the chloride concentration rises at the wells. The infiltrated water replaces the worse quality water and chloride concentrations at the wells are expected to decrease with time.

The nitrate concentration at the locations where the domestic wells are located is comparatively higher than the nitrate concentration below the infiltration pond and the nitrate concentration in the infiltrated water. The nitrate concentration in all domestic wells will be slightly improved.

Social Criteria

In general, the inhabitants are willing to pay more if fully treated wastewater is reused. Respondents do not agree to use the infiltrated water for domestic purposes but they have higher acceptance to use this water for agricultural or industrial purposes. The reuse of treated wastewater for irrigated agriculture would save higher quality groundwater water for drinking water supply and subsequently may solve some environmental problems. The health and religious aspects could be a major concern of people of Gaza to reuse wastewater [13]. The study found that the education level, standard of living and the environment might be key issues in order to convince the people of Gaza to reuse wastewater in agriculture.

Economic Criteria

Implementation of Sc-4 would lead to the maximum benefit. Reuse of wastewater would offer the release of corresponding fresh water resources and will help to expand the overall irrigated area by providing more water to irrigate lands. Hence, the livelihood of the residents may improve. Besides the above-mentioned benefits, more indirect benefits may be gained from improving groundwater quality. These are increased safety and the benefits generated from freeing the land that the current effluent lagoon occupies as well as the other subjective benefits related to seawater intrusion. Finally, the MAR project would create many other supported jobs e.g., related to MAR operation and agricultural activities *etc*.

Strategy Comparison and Ranking

According to the analysis using WLC and PROMETHEE, the same rankings of options were achieved. The comparison of water management options showed that increasing investments in wastewater collection, treatment, and later MAR would result in an improved water management strategy performance with regards to the considered environmental, social, and health criteria. Obvious drawbacks are the investments for infrastructure and their impact on economic feasibility. This should be discussed in greater depth and should be based on comprehensive cost-benefit analysis (CBA) and cost effectiveness that should refer to cost minimization and the related environmental and health benefits, which are fundamental to guarantee the sustainable development of the Gaza Strip.

CONCLUSIONS AND RECOMMENDATIONS

The present study clearly shows the importance of environmental, health, social and economic impact assessment of MAR strategies performing a case study in North Gaza. The integrated approach of combining field campaign, methodological analysis and mathematical modelling has been proven to be effective for a multicriteria decision analysis. In order to increase water supply and to combat water scarcity, water pollution, and health problems at the Northern Gaza Strip, appropriate water resources planning and management measures are urgently required. Reuse of the treated effluent by MAR would strengthen agricultural development and result in increased groundwater availability for domestic and industrial use. The reuse of treated effluent has already been adapted in the national Water Policy for the Gaza Strip [47]. The present study shows that the so-called "Do Nothing Approach" is no real option for Northern Gaza, contributing to further groundwater level decline and groundwater quality deterioration, and increasing health risks for the population of Gaza. The performance analysis of the developed water resources planning and management strategies clearly shows that managed aquifer recharge by infiltration ponds with proper treatment of the effluents is a viable response to the increasing water resources problems of the region. In order to maximize project benefit, optimal pond operation based on practical experiences and regular cleaning of the pond is required to avoid clogging of the pond bed. Application of several MCDA analysis methods probes the robustness of the ranking analysis.

Ten domestic wells will be affected over time due to displacement of relatively low quality groundwater towards the abstraction wells. However, with time, the low quality water will be replaced by the nearby infiltrated water. Special care for water recovery should therefore be planned to protect the existing domestic wells. Another option could be to use the affected domestic wells for agricultural use and use the nearby unaffected wells for domestic water supply. Nevertheless, regular water quality monitoring of abstracted water and efficient recovery wells should be considered. Tremendous effort is required to increase public awareness for wastewater reuse. Adequate water pricing should be made considering the level of income and economic feasibility of the MAR project.

Additional investments should be undertaken for better maintenance and to further extend the wastewater collection network as well as the capacity of the NGWWTP at the Israeli border, accompanying the rapidly increasing wastewater production. Furthermore, managed aquifer recharge contributes to the control of seawater intrusion and groundwater salinity.

Due to the unavailability of scientific data, a variable-density groundwater flow model was not considered in this case study. As the objective of the study is not to quantify salinity intrusion, rather compare different management scenarios, the fresh water flow model is sufficient. In order to investigate the effect of MAR strategies on saline groundwater intrusion into the coastal aquifer, a variable-density groundwater flow model is recommended.

The approach and techniques used in this study can be applicable not only to MAR project implementation but also to other water resource development projects.

ACKNOWLEDGMENTS

The authors gratefully acknowledge the contribution of European Commission (EC) to fund the study under the FP6 program, with the project entitled "Groundwater Artificial recharge Based on Alternative sources of wateR: aDvanced INtegrated technologies and management" (GABARDINE), contract No. 518118. Sincere thanks to the Palestinian Hydrology Group for their helpful support and providing the field data. The authors would also like to express their gratitude to the reviewers and the external editor whose collective input substantially improved this paper. We acknowledge support by Deutsche Forschungsgemeinschaft and Open Access Publishing Fund of Leibniz Universität Hannover.

REFERENCES

1. Maliva, R.G.; Missimer, T.M. Aquifer storage and recovery and managed aquifer recharge using wells: Planning, hydrogeology, design, and operation. In *Methods in Water Resources Evaluation*; Series No. 2; Schlumberger Water Services: Huston, TX, USA, 2010; p. 578.
2. Rahman, M.A. Decision Support for Managed Aquifer Recharge (MAR) Project Planning to Mitigate Water Scarcity Based on Non-Conventional Water Resources. *Ph.D. Thesis*; Department of Applied Geology, University of Goettingen: Goettingen, Germany, 2011. Available online: http://webdoc.sub.gwdg.de/diss/2011/rahman/ (accessed on 22 June 2013).
3. Milgrom, T.; Bensabat, J.; Guttman, J. Towards IWRM implementation—Scenario definition for Kaliya Region. In Proceedings of the 7th International Conference on Water Resources Conservancy and Risk Reduction under Climatic Instability, Limmasol, Cyprus, 25–27 June 2009.
4. Rusteberg, B.; Rahman, M.A.; Bear, J.; Bensabat, J.; Brouyere, S.; Wojda, P.; Sadah, M.A.; Lobo- Ferriera, J.P.; Sauter, M. Decision

support for MAR planning in the context of integrated water resources management. In *Advances in Water Reclamation Technologies for Safe Managed Aquifer Recharge*; Kazner, C., Wintgens, T., Dillon, P., Eds.; IWA Publishing: London, UK, 2012 Chapter 21..

5. Rusteberg, R.; Rahman, M.A.; Sadah, M.A.; Rabi, A.; Sauter, M. Water Management Strategies Analysis Using Multiple Criteria Decision Techniques towards Sustainable Development of Northern Gaza Strip. In Proceedings of BALWOIS 2010: Conference on Water Observation and Information for Decision Support, Ohrid, Republic of Macedonia, 25–29 May 2010.
6. Metcalf & Eddy Inc. *Coastal Aquifer Management Program. Integrated Aquifer Management Program, Task 3*; Report Prepared under USAID Contract No. 294-C-99-00038-00. Metcalf & Eddy Inc.: Gaza, Palestine, 2000; Volume 1.
7. Shomar, B. Groundwater of the Gaza Strip: Is it drinkable? *Environ. Geol.* 2006, *50*, 743–751.
8. Afifi, S. Evaluation of technical performance of Beit-Lahia wastewater treatment plant in the Northern Gaza Strip- Palestine. In Proceedings of the 2nd International Conference on the Palestinian Environment, Nablus, Palestine, 13–15 October 2009; pp. 152–164.
9. Sadah, M.A.; Rabi, A.; Ghannam, S.; Tamimi, A.R. Impact of the "Do-nothing" approach at the Gaza Strip, Palestine. *GABARDINE Deliv. D* 2009, *71*, 21.
10. Engineering and Management Consulting Center (EMCC). *Environmental Assessment: North Gaza Emergency Sewage Treatment Plant Project*; West Bank and Gaza, World Bank: Gaza, Palestine, 2006. Available online: http://www-wds.worldbank.org/external/default/WDSContentServer/WDSP/IB/2006/04/25/000160016_20060425173250/Rendered/PDF/E13870MENA1EA1P074595.pdf (accessed on 2 December 2013).
11. World Bank. North Gaza Emergency sEwage Treatment Additional Financing. Available online: http://www-wds.worldbank.org/external/default/WDSContentServer/WDSP/IB/2008/04/08/000076092_20080411104258/Original/Final0PID.doc (accessed on 23 January 2010).

12. Tubail, K.; Al-Dadah, J.; Yassin, M. Present situation of wastewater and its possible prospect for reuse in the Gaza strip. KA-Abwasser. *Abfall* 2004, *51*, 866–872.
13. Nassar, A.; Tubail, K.; Hall, J. Effluent management of the Gaza central wastewater treatment plant: Assumptions and implications. *J. Environ. Sci.* 2010, *39*, 199–210.
14. Hamdan, S.M.; Nassar, A.; Troeger, U. Impact of Gaza aquifer from recharge with partially treated aquifer. *J. Water Reuse Desalin.* 2011, *1*, 36–43.
15. Al-Agha, M.R. Environmental management in the Gaza strip. *Environ. Impact Assess Rev.* 1997, *17*, 65–76.
16. Shaheen, S.L.R. Nitrate Pollution and Groundwater Modelling of Wastewater Plant in Rafah Area, Gaza Strip, Palestine. *Master's Thesis*, Gent University, Gent, Belgium, 2007.
17. Sheyam, M. Groundwater Salinity Modelling Using Artificial Neural Networks-Gaza Strip Case Study. *Master's Thesis*, The Islamic University, Gaza, Palestine, 2009.
18. Qahman, K.A.; Zhou, Y. Monitoring of seawater intrusion in the Gaza strip, Palestine. In Proceedings of First International Conference on Saltwater Intrusion and Coastal Aquifers Monitoring, Modelling, and Management, Essaouira, Morocco, 23–25 April 2001.
19. Shomar, B. Groundwater contamination and health perspectives in developing world case study: Gaza Strip. *Environ. Geochem. Health* 2011, *33*, 189–202.
20. Aish, A. Hydrogeological Study and Artifitial Recharge Modeling of the Gaza Coastal Aquifer Using GIS and MODFLOW. *Ph.D. Thesis*, Vrije Universiteit Brussel (VUB), Brussels, Belgium, 2004.
21. El-Naeem, M.F.; Heen, Z.A.; Tubail, K. Factors behind groundwater salinization in North governorates of Gaza strip (1994–2004). In Proceedings of 14th International Water Technology Conference (IWTC), Cairo, Egypt, 12–24 March 2010; pp. 893–907.
22. Metcalf and Eddy. *Coastal Aquifer Management Program (CAMP)*; Final Model Report. The United States Agency for International Development and Palestinian Water Authority: Gaza, Palestine, 2000.

23. Shomar, B.; Yahya, A.; Müller, G. Chemical characterization of soil and water from a wastewater treatment plant in Gaza. *Soil Sedim. Contamin. Int. J.* 2005, *14*, 309–332.
24. Palestinian Water Authority (PWA). *Water Resources and Management Issues*; Water Resources and Planning Department: Gaza, Palestinian, 2002.
25. Coastal Municipalities Water Utility (CMWO). Summary about Water and Wastewater Situation in Gaza Strip. Available online: http://www.cmwu.ps/files/reports/8-6-2011-%5Breport%5D-Summary_-Water_and_WW_Situation_in_Gaz_Strip_2010.pdf (accessed on 15 January 2013).
26. Mogheir, Y.; Ajjur, S. Effects of Climate Change on Groundwater Resources (Gaza Strip Case Study). *Int. J. Sustain. Energy Environ.* 2013, *1*, 136–149.
27. Al-Yaqoubi, A.S. Water Resources Statistical Records in Palestine. Available online: http://www.escwa.un.org/divisions/events/Countrypapers/5june07Palestine.pdf (accessed on 11 December 2013).
28. Rahman, M.A.; Rusteberg, R.; Uddin, M.S.; Lutz, A.; Saada, M.A. Site Suitability Mapping and Managed Aquifer Recharge Site Ranking Supported by Mathematical Modelling at Northern Gaza Coastal Aquifer. *J. Environ. Manag.* 2013, *24*, 24–39.
29. Kahan, D. *Agriculture and Water Resources in the West Bank and Gaza, 1967–1987*; Westview Press: Boulder, CO, USA, 1987; p. 181.
30. *Towards Environmental Performance Indicators for the European Union (EU)*; European Commission (EC): Brussels, Belgium, 2002.
31. Organisation for Economic Cooperation and Development (OECD). OECD Environmental Indicators, Development, Measurement and Use. Available online: http://www.oecd.org/dataoecd/7/47/24993546.pdf (accessed on 13 January 2010).
32. Kristensen, P. The DPSIR framework. In Proceedings of Workshop on a Comprehensive/Detailed Assessment of the Vulnerability of Water Resources to Environmental Changes in Africa Using River Basin Approach, Nairobi, Kenya, 27–29 September 2004.

33. Sadah, M.A.; Rabi, A.; Ghannam, S.; Tamimi, A.R. Social considerations regarding the use of brackish water and recovered effluent for agricultural purposes. *GABARDINE Deliv. D* 2009, *74*, 40.

34. Saaty, T.L. *The Analytical Hierarchy Process*; Mcgraw-Hill: New York, NY, USA, 1980.

35. Eastman, J.R.; Kyem, P.A.K.; Toledano, J.; Jin, W. *GIS and Decision Making*; United Nations Institute for Training and Research (UNITAR) European Office: Geneva, Switzerland, 1993.

36. Sadah, M.A.; Rabi, A.; Ghannam, S.; Tamimi, A.R. Environmental impacts of suggested AR response strategies at Gaza Strip, Palestine. *GABARDINE Deliv. D* 2009, *73*, 40.

37. Sadah, M.A.; Rabi, A.; Ghannam, S.; Tamimi, A.R. Economic analysis and feasibility of AR at Gaza Strip, Palestine. *GABARDINE Deliv. D* 2009, *72*, 16.

38. Executive Action Team (EXACT). Temporal trends for water resources data in areas of Israeli, Jordanian, and Palestinian Interest. Available online: http://www.exact-me.org/trends/index.htm (accessed on 11 December 2010).

39. Brans, J.P. Lingenierie de la decision. In *Elaboration Dinstruments Daide a la Decision. Methode PROMETHEE*; Nadeau, R., Landry, M., Eds.; Instruments et Perspectives Davenir, Presses de Universite Laval: Laval, Quebec, Canada, 1982; pp. 183–214.

40. Brans, J.P.; Vincke, P. A preference ranking organization method: The PROMETHEE method for MCDM. *Manag. Sci.* 1985, *31*, 647–656.

41. Brans, J.P.; Mareschal, B. The PROMCALC & GAIA decision support system for multicriteria decision aid. *Decis. Support Syst.* 1994, *12*, 297–310.

42. Macharis, C.; Springael, J.; Brucker, K.D.; Verbeke, A. PROMETHEE and AHP: The design of operational synergies in multicriteria analysis. Strengthening PROMETHEE with ideas of AHP. *Eur. J. Oper. Res.* 2004, *153*, 307–317.

43. Figueira, J.; Greco, S.; Ehrgott, M. *Multiple Criteria Decision Analysis State of the Art Surveys*; Springer Verlag: Berlin, Germany, 2005.

44. Behzadian, M.; Kazemzadeh, R.B.; Albadvi, A.; Aghdasi, M. PROMETHEE: A comprehensive literature review on methodologies and applications. *Eur. J. Oper. Res.* 2010, *200*, 198–215.

45. Abedi, M.; Torabi, S.A.; Norouzi, G.; Hamzeh, M.; Elyasi, G. PROMETHEE II—A knowledge-driven method for copper exploration. *Comput. Geosci.* 2012, *46*, 255–263.

46. California Department of Public Health (CDPH). Groundwater recharge reuse, Draft Regulation. 2008. Available online: http://www.cdph.ca.gov/certlic/drinkingwater/Documents/Recharge/DraftRegulations.pdf (accessed on 3 February 2010).

47. Nassar, R.A.; Al-Najar, H.; Al-Dadah, J.Y. Socio-economic aspects of wastewater reuse in the Gaza strip. *J. Environ. Sci. Technol.* 2009, *2*, 170–178.

chapter 5

Application of Hydrologic Tools and Monitoring to Support Managed Aquifer Recharge Decision Making in the Upper San Pedro River, Arizona, USA

Laurel J. Lacher [1], Dale S. Turner [2], Bruce Gungle [3], Brooke M. Bushman [2] and Holly E. Richter [2]

[1]Lacher Hydrological Consulting, 821 S. Meyer Avenue, Tucson, AZ 85701, USA

[2]The Nature Conservancy, 1510 E. Fort Lowell Road, Tucson, AZ 85719, USA

[3]Arizona Water Science Center—U.S. Geological Survey, 520 N Park Avenue, Tucson, AZ 85719, USA

ABSTRACT

The San Pedro River originates in Sonora, Mexico, and flows north through Arizona, USA, to its confluence with the Gila River. The 92-km Upper San Pedro River is characterized by interrupted perennial flow, and serves as a vital wildlife corridor through this semiarid to arid region. Over the past century, groundwater pumping in this bi-national basin has depleted baseflows in the river. In 2007, the United States Geological Survey published the most recent groundwater model of the basin. This model served as the basis for predictive simulations, including maps of stream flow capture due to pumping and of stream flow restoration due to managed aquifer recharge. Simulation results show that ramping up near-stream recharge, as needed, to compensate for downward pumping-related stress on the water table, could sustain baseflows in the Upper San Pedro River at or above 2003 levels until the year 2100 with less than 4.7 million cubic meters per year (MCM/yr). Wet-dry mapping of the river over a period of 15 years developed a body of empirical evidence which, when combined with the simulation tools, provided powerful technical support to decision makers struggling to manage aquifer recharge to support baseflows in the river while also accommodating the economic needs of the basin.

INTRODUCTION

Social Context

Balancing the social and economic water needs of humans with those of the environment, frequently referred to as sustainable water management, presents a global challenge, especially in arid and semi-arid regions, such as the American Desert Southwest. Understanding that the concept of sustainable water management is subjective, particularly in areas where human groundwater extractions have long exceeded average annual natural recharge, this study focuses on the tension between human demands for groundwater and the groundwater needs of an exceedingly rare perennial river system within the same basin of Southeast Arizona.

We report here on a combination of hydrologic tools and monitoring that are supporting significant progress toward many measures of sustainability on the scale of decades at one important site. The case for developing a regional groundwater recharge network along the Upper San Pedro River is built upon a series of studies and reports funded by the Upper San Pedro Partnership (Partnership), and vetted by its Technical Committee over a period of approximately 15 years. The Partnership is a consortium of 23 agencies and organizations working together to meet the long-term water needs of the Sierra Vista Subwatershed by achieving sustainable yield of the regional aquifer to: (1) preserve the San Pedro Riparian National Conservation Area (SPRNCA); and (2) ensure the long-term viability of Fort Huachuca. The purpose of the Partnership is to coordinate and cooperate in the identification, prioritization and implementation of comprehensive policies and projects to assist in meeting water needs in the Sierra Vista Subwatershed of the Upper San Pedro River Basin. Richter and others [1] provide a more complete discussion of the purpose, methods, and vision encompassed by the Partnership. In addition, recent groundwater modeling efforts have focused stakeholders' collective understanding of the hydrologic system on actionable strategies. Together, this body of work helps identify potential spatial and temporal groundwater recharge targets for mitigating many of the negative environmental, regulatory and economic consequences that may result if groundwater inputs to the San Pedro River diminish as a result of pumping, as predicted by groundwater modeling. Using this information, three of the stakeholders including the U.S. Army Compatible Use Buffer (ACUB) Program, Cochise County, and The Nature Conservancy are collaborating to develop an aquifer protection and recharge network of sites in the San Pedro River basin to protect flows in the river from anticipated pumping-related depletions over the next 50 to 100 years. This interim solution is intended to allow time to develop other longer-term strategies for addressing issues of the accumulated groundwater deficit in the regional aquifer and climate change. Examples of such longer-term strategies could include gradual elimination of consumptive use of groundwater, limitations on new pumping, enhanced utilization of urban runoff, and importation of water from outside the basin. Most of these strategies are controversial and/or expensive, and would require significant political and legal efforts to secure physical water supplies. This approach—using the best available science to implement a relatively uncontroversial and legally available interim solution to preserve baseflows while more difficult long-term solutions are pursued—may serve as a model for other dryland river basins.

Study Area

The San Pedro River flows north 279 kilometers (km) from its headwaters in northeastern Sonora, Mexico, to its confluence with the Gila River in southeastern Arizona, U.S.A. This paper focuses on the 2460-km^2 Sierra Vista Subwatershed ("subwatershed") of the Upper San Pedro basin ("basin"), north of the United States-Mexico boundary (Figure 1). This subwatershed area includes about 47 km of the river, most of it within the San Pedro Riparian National Conservation Area (SPRNCA), designated by the United States Congress in 1988 to protect and enhance the riparian area and its aquatic resources [2].

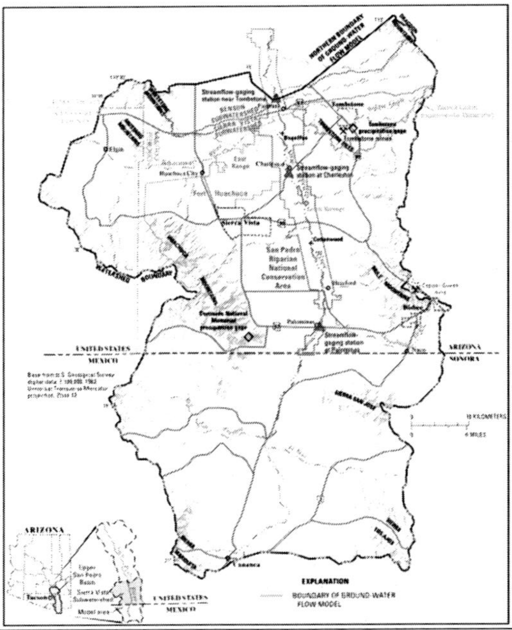

Figure 1. Map of the Upper San Pedro Basin showing the extent of the USGS groundwater flow model [3] and the San Pedro Riparian National Conser-

vation Area. The Sierra Vista subwatershed is the area north of the United States—Mexico boundary.

Elevations within the Sierra Vista subwatershed range from more than 2800 meters (m) above mean sea level in the Huachuca Mountains on the western edge of the basin, to 1052 m at the Tombstone stream-flow gaging station at the north (downstream) end of the subwatershed. Precipitation across the topographically diverse basin ranges from 35 centimeters per year (cm/yr) near Tombstone in the central-north part of the subwatershed to about 76 cm/year in the highest parts of the Huachuca Mountains [4,5]. Like other areas in the desert southwest, precipitation in the study area is predictably bimodal with a summer monsoon season from July through mid-September accounting for about one half, and a winter wet season from December through March that accounts for another one third of the annual precipitation. Occasionally, tropical storms will trigger very large runoff events in October, but otherwise the fall months are typically dry. Stream flow is lowest in May and June, with peak monthly flows occurring during the summer monsoon season (Figure 2). Average monthly minimum flows range from about 1300 cubic meters per day (cu-m/d) at the Palominas stream-flow gaging station in the south (upstream) end of the subwatershed to 19,300 cu-m/d at the Charleston station, and 8100 cu-m/d at the Tombstone station on the north (downstream) end of the study area (Figure 2). Snow melt in the Huachuca Mountains often supports flow in mountain springs and intermittent streams in the spring months of March and April.

Groundwater discharge from the regional basin-fill aquifer supports baseflows in the river and evapotranspiration by near-stream phreatophytes. Groundwater pumping from that same aquifer supplies a human population in the region that is expected to increase by 46% by the year 2050 [6]. Sierra Vista, Arizona, is the largest municipality in the basin with a population of roughly 45,300 [7]. The U.S. Department of Defense installation at Fort Huachuca borders Sierra Vista on the west and north sides and is a major economic force in the basin [8].

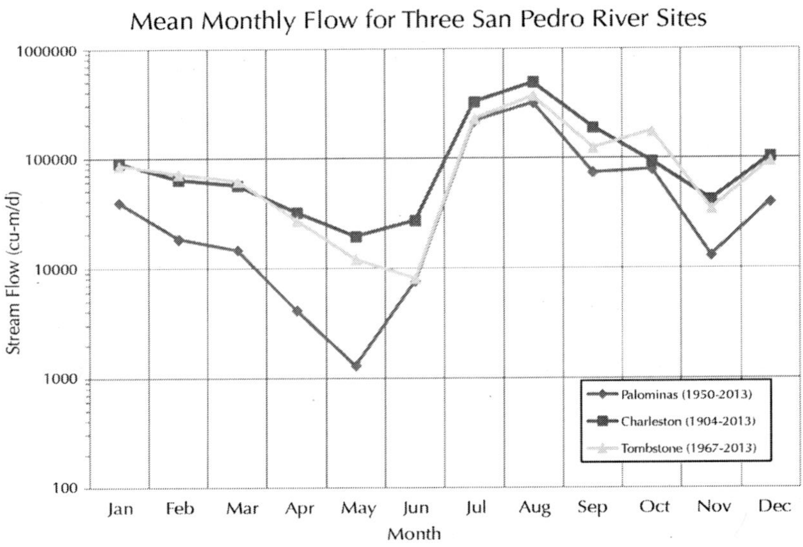

Figure 2. Mean monthly stream flow for the three long-term monitoring sites (shown in Figure 1) on the Upper San Pedro River in the Sierra Vista subwatershed for the period of record at each site. The y-axis is plotted on a log scale to highlight the minimum flows in May and June.

Because of its federal designation and its globally significant biodiversity [9], the SPRNCA has been the subject of extensive hydrological and ecological research. One major finding was that stream flow permanence explained most variance in the basal area of two keystone riparian species, Fremont cottonwood (*Populus fremontii*) and Goodding willow (*Salix gooddingii*) [10,11]. As a result, conservation strategies have focused on maintaining or increasing stream flow through reductions in groundwater pumping and managed aquifer recharge at key locations.

Hydrological Studies in the Basin

General Hydrologic Processes

The Upper San Pedro River is one of the most intensively studied semi-arid stream systems in the world. Estimated ages of the earliest paleo-Indian sites along the banks of the ancestral San Pedro River date back to 11,000 to 13,000 years ago [12,13]. Two Clovis Indian sites have been associated with mammoth kills in the subwatershed, providing some of the earliest dates for human hearths of this culture [13]. The basin has been grazed since the time of the first Spanish explorers in the 1500s [14], and has hosted two major metals mining operations in Tombstone (late 1800s) and Bisbee (1880s–1975), Arizona. In addition, Cananea, Mexico boasts a major copper mine that is currently in active production. The major U.S. Department of Defense Army installation at Fort Huachuca has been in active operation since 1877 and has one of the earliest water rights claims in the basin [15]. More recently, the town of Sierra Vista has grown to include over 45,000 people (including Fort Huachuca). All of these activities have had significant impacts on the landscape and water resources in the basin. Hereford [14] presents a clear description of the land use changes coupled with exceptional periods of flooding near 1900 that led to significant entrenchment and stabilization of the San Pedro River stream channel and dewatering of much of the shallow alluvium of the river channel.

The Walnut Gulch Experimental Station (Figure 1) was established in the northeast area of the subwatershed by the Research Division of the Soil Conservation Service in 1951, and has been the source of continuous instrument-based hydrologic and rangeland studies since that time [16]. In 1966, Brown and others [17] evaluated the water resources of Fort Huachuca, which lies immediately west and north of the City of Sierra Vista (Figure 1). Brown and Aldridge [18] estimated San Pedro River surface discharge from the international boundary with Mexico to its confluence with the Gila River and inputs to the system from tributary inflow and from mountain-front recharge. Much of the assessment of hydrologic resources in the basin that followed came as a byproduct of the development of groundwater models, including those by Freethey [19], Vionnet and Maddock [20], Corell and others [21], Goode and Maddock [22], and Pool and Dickinson [3].

The Arizona Department of Water Resources (ADWR) evaluated the groundwater resources of the basin in 1990 [23], and then again in 2005 [24], in order to determine whether or not the basin should be classified as an Active Management Area (AMA) under the 1980 Groundwater Management Act of Arizona. The ADWR determined that the basin did not meet the statutory criteria for AMA designation [23,24], but much of the research that was conducted as part of that assessment remains highly relevant. Pool and Coes [25] described the state of the knowledge of hydrogeology of the subwatershed. As part of that work, an extensive monitoring program was initiated in the subwatershed that included geophysical surveys, ephemeral stream flow monitoring, installation or refurbishment of a number of stream gaging stations, aquifer storage monitoring using microgravity techniques, and basin-wide water level monitoring. Much of this monitoring program continues today.

In support of the conservation mission of the SPRNCA, Section 321 of the Defense Authorization Act of 2004 [26] revised how the federal Endangered Species Act applies to the Fort Huachuca Military Reservation and directed the Partnership to "...restore and maintain the sustainable yield of the regional aquifer [of the Sierra Vista Subwatershed] by and after September 30, 2011." It also required annual progress reports on these efforts, which were produced for most calendar years from 2002 through 2011 (see [27], for example).

The Partnership sponsored several research reports during this time, which provided the scientific basis for development of the U.S. Geological Survey (USGS) groundwater model by Pool and Dickinson [3]. These studies include Coes and Pool's [28] assessment of ephemeral-stream channel and basin-floor infiltration and recharge, Gungle's [29] analysis of the timing and duration of ephemeral stream flow in the subwatershed, and Leenhouts and others' [30] analysis of the hydrology, vegetation-hydrologic relationships, and evapotranspiration requirements and plant-water sources in the SPRNCA. Leenhouts and others [30] established additional streamflow and groundwater monitoring in the subwatershed, including streamflow stage and permanence data, near-stream alluvial aquifer groundwater and vertical gradient monitoring, and a continuous meteorological and eddy covariance monitoring station for measurement of evapotranspiration in the SPRNCA.

A statistical analysis of the trends in streamflow in the San Pedro River was also published in 2006 by Thomas and Pool [5]. Leake, Pool and Leenhouts [31] used Pool and Dickinson's [3] five-layer groundwater model of the basin to conduct a capture and recharge analysis that mapped the effects of pumping and recharge across the subwatershed on groundwater discharge to the SPRNCA. Kennedy and Gungle [32] analyzed baseflow discharge from the subwatershed at the USGS gaging station near Tombstone, Arizona, at the north end of the subwatershed. Most recently, Lacher [33] updated the Pool and Dickinson [3] groundwater flow model to include recent changes in pumping and artificial recharge in the subwatershed, and simulated the projected effects of population growth-driven increases in pumping on groundwater levels and baseflow through 2105.

Managed Aquifer Recharge

In 2006, Stantec [34] developed a Flood Control Urban Runoff Plan for Cochise County that evaluated the size, placement, and efficacy of 30 existing and proposed stormwater detention basins on the west side of the subwatershed functioning as de facto recharge basins. As part of this study, GeoSystems Analysis [35] used a detailed precipitation-recharge-stormwater runoff regression model, based on the U.S. Department of Agriculture's Automated Geospatial Watershed Assessment Tool [36] for the urbanized Coyote Wash watershed in Sierra Vista to estimate the cost, recharge volume, and urban-enhanced runoff for that watershed and several others on the west side of the San Pedro River within the subwatershed. GeoSystems Analysis previously developed the stormwater regression model used for the Upper San Pedro Partnership and Cochise County based on detailed AGWA stormwater modeling results for the Coyote Wash watershed located in the City of Sierra Vista [37].

Predictive groundwater modeling [33] indicates that within the next 100 years, two regional cones of depression will merge and reduce groundwater flow from the regional aquifer to the San Pedro and Babocomari (a major tributary to the San Pedro) rivers (Figure 3). These simulations incorporate pumping increases over time that reflect U.S. Census growth rates [5,6] and maintain constant natural recharge and evapotranspiration at 2003 levels (as published in the USGS groundwater model [2]) for the entire 21st century. Figure 4a maps the west-east transect A-A′ for the simulated groundwater-level profiles shown in Figure 4b. This transect goes through the Sierra Vista-Fort Huachuca cone of depression, and illustrates how simulated groundwater levels at the groundwater divide between the cone of depression and the river have already declined by 17% since pre-development conditions in 1902 and are predicted to double that level of decline, to 35%, by the year 2100. These reductions in groundwater levels produce commensurate percent reductions in groundwater gradient (change in head divided by distance) as measured from the same point on the groundwater divide to the river. While gradients directly under the river are not yet showing the same degree of impacts as those farther west, the ultimate outcome of these simulated changes would be greater baseflow losses in some losing reaches, smaller gains in some gaining reaches, and possibly the conversion of some reaches from gaining to losing. The resulting groundwater-driven reduction in baseflow and shallow groundwater would likely impair the dependent riparian systems [33].

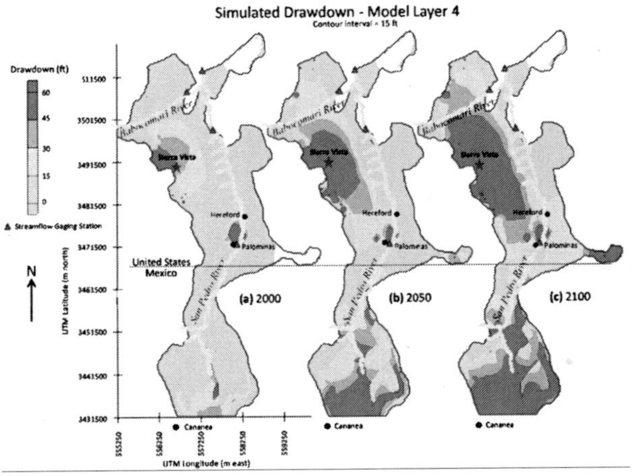

Figure 3. Simulated drawdown (m) in the primary regional aquifer of the basin (a) in 2000; (b) in 2050; and (c) in 2100 [33].

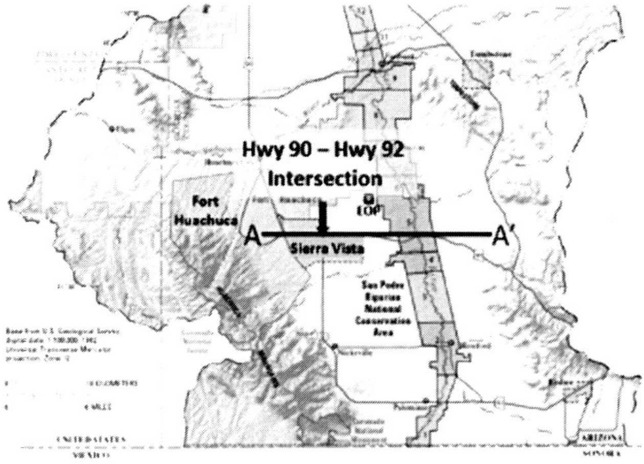

(a)

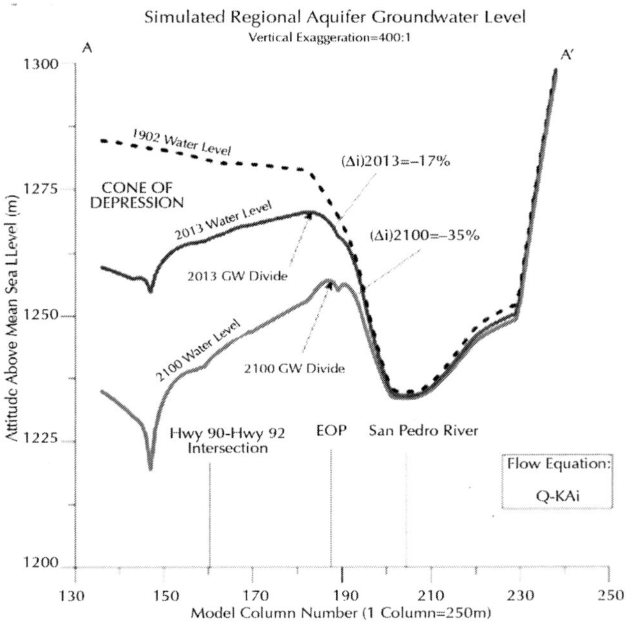

(b)

Figure 4. (a) Cross-section A-A′ in the center of the Sierra Vista subwatershed (adapted from Stromberg and others [10]; (b) Simulated groundwater-level profiles in 1902, 2013, and 2100 along the A-A′ cross-section showing change in gradient (Δi) from −17% in 2013 to −35% in 2100.

One potential solution to the problem of aquifer storage depletion would entail importation of at least some water from a source outside the basin [38]. However, at this time, importation does not appear likely in the foreseeable future (see companion paper by Richter and others [1]). In lieu of a replacement water supply, water managers and stakeholders in the subwatershed recognize the need for a near- to medium-term (*i.e.*, years to decades) intervention to protect stream baseflows and the associated riparian areas.

The City of Sierra Vista has more than 10 years of monitoring data from its Environmental Operations Park (EOP) where treated effluent is recharged at a rate of roughly 9100 cubic-meters per day (m^3/d) through artificial wetlands and recharge basins between the City's pumping center and the San Pedro River. This project was designed and constructed to create a groundwater mound to sustain surface water flow and supplement alluvial groundwater levels during low-flow periods [39]. Groundwater modeling that incorporates the EOP recharge data indicates that the project is successfully contributing recharge to both the nearby cone of depression in the regional aquifer and to baseflows in the San Pedro River [33,40]. Further groundwater modeling presented in this study suggests that additional recharge projects designed to utilize enhanced urban runoff, treated effluent or other local supplies near the San Pedro and Babocomari rivers may successfully mitigate anticipated pumping-related baseflow depletions for up to 100 years.

While conservation efforts within the subwatershed have reduced consumptive use of groundwater in the basin, particularly on the Fort Huachuca Army installation, the cumulative storage loss in the regional aquifer within the subwatershed from the past half century of pumping, estimated to exceed 800 MCM (refer to explanation provided in Section 2.3.2) remains a threat to the San Pedro River and its tributaries by intercepting mountain-front recharge and by reducing the groundwater gradient that drives aquifer discharge to the rivers. Because of this storage deficit and the fact that virtually all water use in the subwatershed depends on groundwater, almost any level of continued groundwater use poses an increasing risk of stream and riparian evapotranspiration capture over time.

Streamflow Permanence Monitoring

Direct measurement of baseflow, defined as the portion of streamflow derived from groundwater, is complicated by the effects of prolonged storm and/or snowmelt runoff (which tend to exaggerate baseflow estimates) and evapotranspiration (which reduces apparent baseflow). Acknowledging these complexities, practitioners have found that inter-annual stream-flow permanence in a river system that experiences intermittent drying in some reaches is a useful indicator of both hydrologic and ecological conditions in the San Pedro basin. Using a technique called wet/dry mapping, analysis of surface water presence (strict criteria are used to ensure consistent definition of surface water versus minimal puddles) during the driest time of year (mid June in the San Pedro Basin) reveals areas with high year-to-year variation in the length of surface wetting (Figure 5). By limiting the influence of storm events as much as possible, and assuming no significant changes in the condition of the riparian forest, these variations in wetted length are believed to be the best available physical expression of local groundwater conditions and may provide early warning of ecological changes [41].

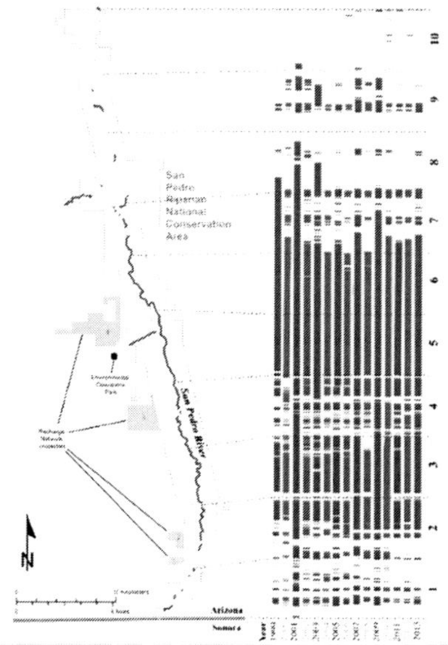

Figure 5. Wet/dry results from the San Pedro Riparian National Conservation Area. The heavy river line shows reaches, which were wet in June 2013.

Bars on right side represent wet reaches for each year, 1999–2013. Labels on the far right identify the 10 analysis segments, each covering 8.1 km. The four green properties (parcels 1 through 4) on map were recently acquired for groundwater protection and recharge projects. Redrawn from Turner and Richter [41].

Wet/dry mapping, as applied in the San Pedro River basin, uses citizen scientists annually to map the spatial extent of surface water in a river or stream. The method provides a comprehensive snapshot of conditions for the whole river at the same date each year, allowing comparisons of year-to-year variability [41]. Beginning in 1999, staff from The Nature Conservancy and the U.S. Bureau of Land Management coordinated volunteers to map the spatial extent of surface water within the SPRNCA. The exact dates varied slightly, but mapping was conducted during the third weekend of June each year to coincide with the lowest flow before the expected start of the summer rainy season. Through the years, progressively more of the river and its tributaries have been surveyed, increasing to 231 km of the mainstem and 266 km of tributaries in 2013. This paper addresses only the 80 km of mainstem river surveyed through SPRNCA.

METHODS

Wet/Dry Mapping

Surveyors walked predetermined segments of the river, recording the coordinates of beginning and end points of all surface water segments greater than or equal to 9.1 m long using paper data forms and consumer-grade Global Positioning System units. They disregarded any dry gaps less than 9.1 m long in otherwise wet reaches. The resulting point coordinates were imported to a Geographic Information System (ArcGIS, Environmental Systems Research Institute, Redlands, CA, USA), and snapped to the closest points on a linear representation of the river. To identify localized trends, organizers partitioned the SPRNCA into 10 equal segments, 8.1 km long. Segments were analyzed for probability of trend using the Mann-Kendall test. For graphic purposes, we calculated and display the Sen estimate of linear trend (detailed methods are provided by Turner and Richter [41]). Starting in 2007, maps and summary data from the wet/dry surveys have been posted each year to a web site (http://www.azconservation.org) for public distribution.

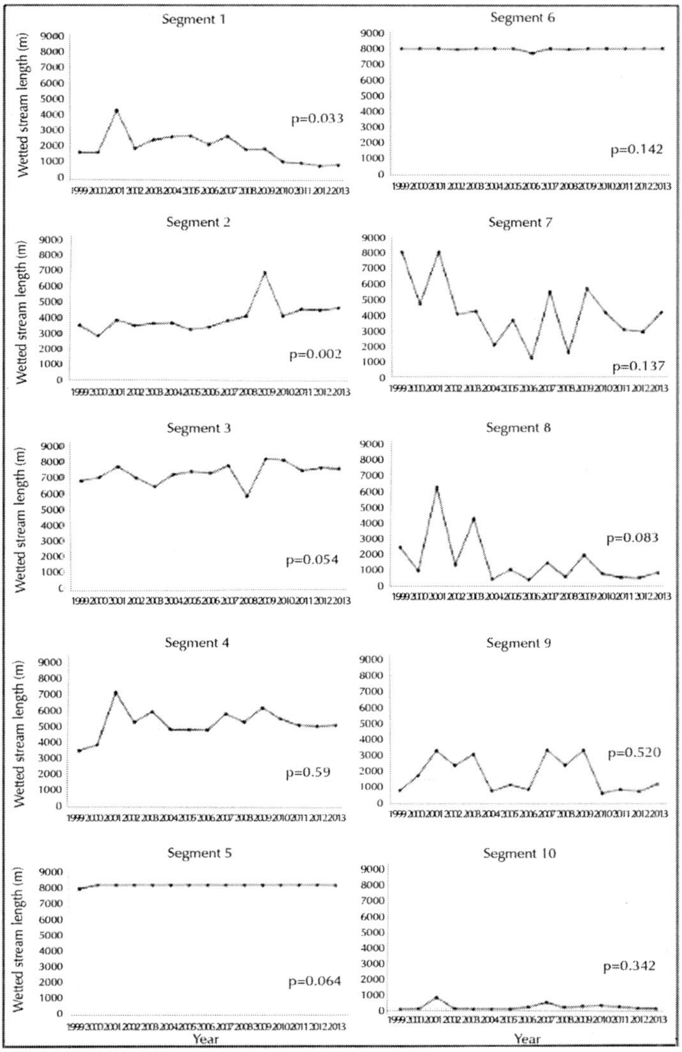

Figure 6. Total wetted lengths for the 10 analysis segments in Figure 3. Segment numbers increase from south to north (downstream). Revised from Turner and Richter [41].

Results from wet/dry mapping in the SPRNCA (Figure 6), as a whole, show about half the river has permanent surface water with some year-to-year variation but no trends. However, analysis of the smaller segments shows considerable variation Turner and Richter [41]). The southernmost segment, Segment 1, displays a significant downward trend, while Segment 2 trends upward. Most of the other segments have either year-to-year up/down results or a stable condition without trend. The distribution of wet and dry reaches has provided a simple way to prioritize and site conservation actions aimed at reduced groundwater pumping and managed aquifer recharge, as described in Section 3 below. Wet/dry mapping data are also expected to provide a quantitative measure of conservation progress after those strategies are implemented.

CAPTURE MAPPING

Overview of Capture

The Upper San Pedro River's position near the geographic center of an alluvial basin makes it a classic example of basin-fill hydrology. The existence of a major pumping center at Sierra Vista/Fort Huachuca between the primary area of recharge for the subwatershed (the Huachuca Mountains on the western boundary) and the river (Figure 1) also makes the basin a classic case study in groundwater capture. Capture is the increase in recharge to, and (or) decrease in discharge from, a basin that eventually occurs as a result of groundwater pumping. Theis [42] first addressed the consequences of groundwater pumping from a previously undeveloped system in 1940:

> "Under natural conditions…previous to development by wells, aquifers are in a state of approximate dynamic equilibrium. Discharge by wells is thus a new discharge superimposed upon a previously stable system, and it must be balanced by an increase in the recharge of the aquifer, or by a decrease in the old natural discharge, or by loss of storage in the aquifer, or by a combination of these."

Lohman and others [43] further clarified the definition of capture from Theis' description:

"The decrease in discharge plus the increase in recharge is termed capture."

This definition of capture may be written as:

$$\text{Capture} = \Delta R + \Delta D \qquad (1)$$

where ΔR and ΔD equal the increase in recharge and the decrease in discharge, respectively.

In the Upper San Pedro basin, as in many basins in the southwest deserts of the United States, aquifer storage is plentiful as a result of recharge over thousands of years of depositional history, but potential sources of capture are limited primarily to reductions in riparian ET, reductions of groundwater discharge to streams (baseflow), and direct capture of streamflow. The numerical partitioning of extracted groundwater into its constituent sources through a water-budget process is a common and fraught practice in many water-short areas of the United States. Bredehoft, Papodopulos, and Cooper [44] attributed this practice to "perhaps the most common misconception in groundwater hydrology" which is "that a water budget of an area determines the magnitude of possible groundwater development." Under this line of reasoning, many water managers have concluded that all water entering the system as natural recharge is available for extraction without long-term deleterious effects. Brown [45] addressed the problem with this argument through an example of a well whose cone of depression eventually expands to intersect a stream in which he demonstrated that,

"...the rate at which the hydrologic system reaches a new steady state depends on the rate at which the natural discharge [in his example to a stream] can be captured by the cone of depression."

Figure 7 illustrates the evolution of a groundwater system which receives natural recharge (through precipitation) at a fixed and limited rate, R_0, and discharges to a stream as baseflow at a rate of D_0. In Figure 7a, the system is in equilibrium prior to any significant groundwater development. In this natural state of equilibrium (roughly prior to 1940 in the Upper San Pedro basin [3]) recharge equals discharge (Equation (2)):

$$R_0 = D_0 \qquad (2)$$

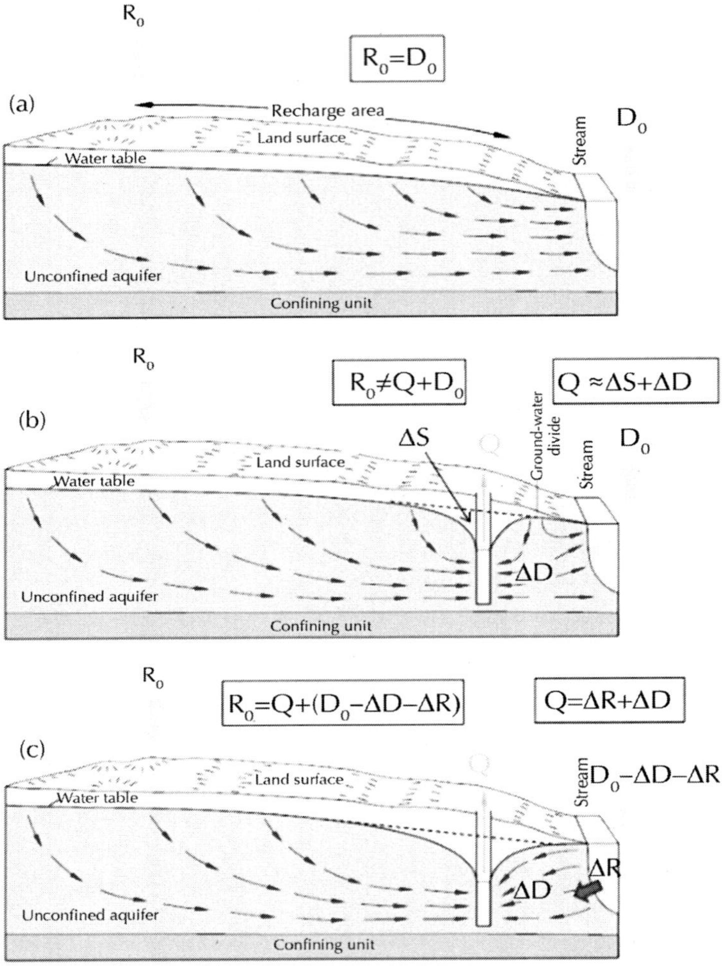

Figure 7. (a) Groundwater discharging to stream under equilibrium conditions; (b) Pumping at rate Q from a well intercepting groundwater that would have discharged to the stream under equilibrium conditions; Q is roughly equal to the rate of change in aquifer storage (ΔS) plus reduced groundwater discharge to the stream (ΔS); (c) Pumping at the same rate (Q) under a new equilibrium condition. Now, pumping is reducing groundwater discharge to the stream (ΔD) and inducing recharge directly from the stream (ΔR), and aquifer storage depletion ceases (ΔS = 0). Adapted from Winter and others ([46], Box C, p. 15) and from Heath ([47], p. 33).

In Figure 7b, pumping at a rate of Q is superimposed on the system, producing a rapid rate of change in aquifer storage near the well (**Δ***S*) and intercepting some groundwater that would otherwise have discharged to the river (**Δ***D*), but having no immediate effect on discharge to the stream (D_0) due to the persistence of a groundwater "high" between the well and the river. At this point, pumping is roughly equivalent to the rate of change in aquifer storage plus the reduction in groundwater discharge to the stream (Equation (3)).

$$Q=\Delta S+\Delta D \tag{3}$$

and recharge and discharge are no longer in balance (Equation (4)):

$$R_{0\neq Q+\ D0} \tag{4}$$

After some time (Figure 7c), the cone of depression intercepts the stream and reverses the gradient of the groundwater so that it now flows toward the well from all directions, directly capturing streamflow. Even though the pumping rate (Q) remains constant, recharge from precipitation also remains fixed at R_0, so the only sources of water for extraction come reduced groundwater discharge to the stream, capture (increased recharge) from the stream, itself, and possibly some additional aquifer storage. A new equilibrium is achieved when aquifer storage is no longer being depleted (**Δ***S* = 0) and pumping is balanced by the capture of stream flow and decreased discharge to the stream (Equation (5)):

$$R_{0=Q+(D0-\Delta}D-\Delta R) \tag{5}$$

or,

$$Q=\Delta R+\Delta D \tag{6}$$

Lohman [48] identified another potential source of capture not expressly described above. He referred to it as "salvaged rejected recharge from precipitation." This potential source of capture would comprise a new source of recharge (*ΔR*) as described in Equations (1) and (6). In the case of a large alluvial basin like the San Pedro, where much of the shallow alluvial aquifer is separated by a thick sequence of confining materials from the underlying regional aquifer, this source of potential capture would derive from the occasional replenishment of the shallow alluvial aquifer during runoff from one or more very large precipitation events. Runoff exceeding 25M cu-m/d occurs roughly every 5 to 6 years at the Palominas station (Figure 1), so any excess alluvial aquifer storage opened up by groundwater pumping could capture flood flows that would otherwise have been rejected and remained in the stream.

For the period 2000 to 2009, Kennedy and Gungle [32] found that alluvial aquifer storage changes that govern baseflow measured at the Tombstone stream-flow gaging station (near the far downstream end of the study area (Figure 1)) were a function of upstream riparian ET and summer precipitation. Although they did not identify groundwater pumping as a clear source of baseflow decline in that reach over that short period, they do caution that, "[c]ontinued regional groundwater pumping will, however, eventually lead to a decline in the contribution of regional groundwater to base flow." This contrast between the interannual scale of fluctuations in alluvial aquifer storage and the multi-decadal scale of changes in regional aquifer discharge is highly significant for long-term water management planning. Our study focuses on these longer-term regional aquifer changes, but acknowledges the importance in improving our understanding of the linkages between sources of capture at different temporal and spatial scales.

Groundwater simulations [3,33] indicate that the Upper San Pedro basin is transitioning between the second and third scenarios illustrated in Figure 7. Of course, streamflow is not the only potential source of capture in most groundwater basins, and may not be available at all, as in basins where the streambed has lost connectivity with the underlying aquifer. Other potential sources of capture include riparian evapotranspiration, groundwater inflow from boundary sources, and groundwater outflow from the basin. Because groundwater and surface water systems operate on such different time scales, stresses on a groundwater system may not manifest as baseflow depletions for many years. Unfortunately, "in many circumstances the dynamics of the groundwater system are such that long periods of time are necessary before any kind of an equilibrium condition can develop. In some circumstances the system response is so slow that [groundwater] mining will continue well beyond any reasonable planning period" ([44], p. 55–57). The goal of the study described in this paper is to anticipate the future impacts of 20th and 21st century pumping on the San Pedro River and to develop a strategy to mitigate those effects as they occur.

Development of Capture Map

Leake and others [31,48] developed a unique tool that utilizes the Pool and Dickinson [3] regional groundwater flow model to assess simulated pumping-induced capture of streamflow, spring discharge, and riparian ET across the model domain in response to a unit pumping stress at every location in the model, and presents the results as a map overlay with colored contours representing the amount of capture, as a fraction of the pumping rate, after pumping for a specified amount of time (Figure 8). Leake and others [49] provide detailed steps for developing a capture map using a groundwater model. Alternatively, capture for a given location can be calculated as the pumping time needed to reach a depletion-dominated supply (the time at which capture begins to provide greater than 50% of total groundwater pumped). A typical time scale for a basin in the American Southwest might be 0 to 100 years [50]. Capture map development can also be run in reverse, providing estimates of the total increase in streamflow, riparian evapotranspiration, and spring flow, as a fraction of recharge rate, after recharging at a unit rate for a specified amount of time [51]. The two are not necessarily the inverse of each other. Leake and others [31] calculated stream and riparian area capture resulting from pumping in the lower basin fill primary aquifer (layer 4 of the 5-layer groundwater model [3]) in the Upper San Pedro basin. They used the same method to calculate response in the stream/riparian area from recharge applied to the top-most layer of the model that overlies the extent of layer 4 (layers 1, 2, or 4).

Use of Capture Map in the San Pedro Basin

Richter and others [1] discuss the evolution of the capture map concept and its use in policy development within the basin. While the capture map does not replace groundwater modeling, it is a simple, intuitive tool that permits the layman to gain a better understanding of the degree of connectivity between the groundwater system and the river. Policy makers and stakeholders embraced the capture map in the San Pedro basin as a guide for various preliminary decisions on community development and ecological restoration.

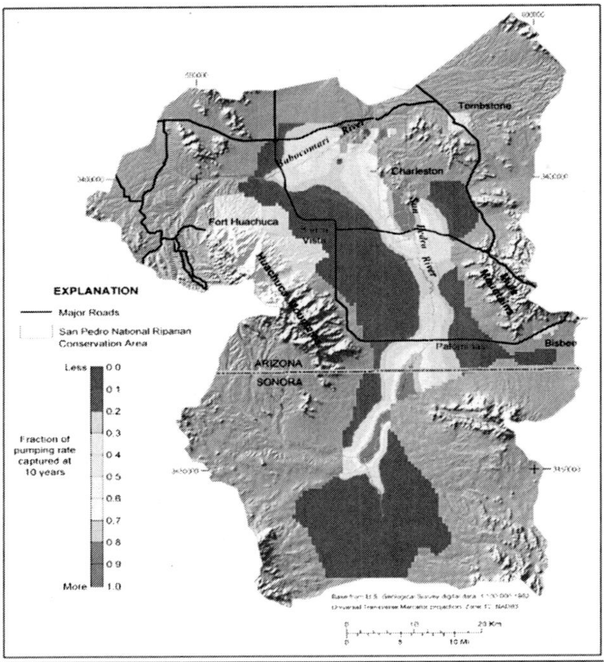

Figure 8. Computed capture (as a percentage of pumping rate) of streamflow, riparian evapotranspiration, and spring flow that would result for withdrawal of water from model layer 4 at a constant rate for 10 years. The color at any location represents the fraction of the withdrawal rate by a well at that location that can be accounted for as changes in outflow from and or inflow to the aquifer for model boundaries representing streams, riparian vegetation, and springs. Redrawn from Leake and others [31].

Advantages/Disadvantages and Limitations of the Tool

Capture maps highlight where pumping will have the greatest impact on a water resource (such as a river) within a specified period of time, and thus make an excellent preliminary planning tool for water managers, commercial and residential developers, and conservationists who can benefit from reducing the immediate impacts of groundwater pumping. The similarly constructed recharge map provides similar planning benefits, offering the planner an initial overview of the most advantageous sites for storm-water, treated effluent, or other recharge facilities.

Despite the value of capture maps for communicating some complex hydrologic concepts to lay audiences, details about specific volumes and rates of capture for specific sources/sinks in the basin cannot be extrapolated directly from this tool. Capture maps, based on linear superposition, reflect the assumption of constant hydrologic properties of the aquifer, assuming that pumping causes no non-linear behavior in the hydrologic system. This limitation, and the fact that each cell of the model is stressed in isolation to make the map, means that capture maps cannot replace the use of a full groundwater model for examining the cumulative impacts of various stresses and sinks that change over time or that cause nonlinearities in hydrologic properties or boundary conditions.

NEAR-STREAM RECHARGE SIMULATIONS

Description of Model

Simulations in this study used the most recent and most sophisticated groundwater flow model of the Upper San Pedro Basin available [3]. Although the model area includes all of the 4500-km^2 basin of which 40% is in Mexico, this study focuses on the Sierra Vista subwatershed within the United States (Figure 1). This MODFLOW-2000 [52] model is based on a uniform 250 m × 250 m grid spacing oriented north-south in alignment with the basin. The stream is represented by the Stream Package [53], hydraulic flow is modeled with the Layer Property Flow (LPF) package, and riparian evapotranspiration is modeled with the Evapotranspiration (EVT) Package of MODFLOW-2000. The two-season model reflects the seasonal significance of evapotranspiration in the riparian zones along the San Pedro and its tributaries. The cool season extends from mid-October to mid-March, and the warm season runs from mid-March through mid-October [3].

As with all groundwater models, this model has several limitations. Most significantly, it does not simulate flood flows, which are known to contribute significant recharge to the alluvial aquifer near the center of the basin as well as along some ephemeral tributaries to the mainstem of the San Pedro River. This seasonal "topping off" of the shallow alluvial aquifer may support baseflows in the river through one or more dry seasons, and is an important component of the riparian system. While this shortcoming means that the model tends to underestimate true baseflow (regional aquifer plus shallow alluvial aquifer contributions), it does not preclude the model as a useful tool for analyzing the effects of pumping on the regional aquifer's contribution to baseflow. For this reason, the term "baseflow" in this study refers strictly to that component of total baseflow that derives from the regional aquifer where most of the pumping in the basin occurs.

Figure 9 illustrates the conceptualized hydrogeologic cross section (upstream view) of the subwatershed and shows how the model layers correspond to that conceptualization. The model structure includes five layers in a stacked-bowl configuration representing sediment accumulation in the structural depression between two bounding mountain ranges, as is typical of the Basin and Range province of the western United States [3]. Only model layer 5 is found throughout the entire region (Figure 10).

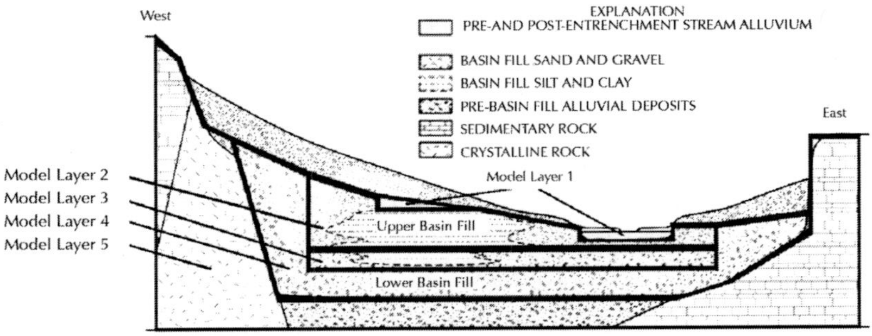

Figure 9. Conceptualized cross section of basin showing model layers. Adapted from Figure 3 in Pool and Dickinson [3]).

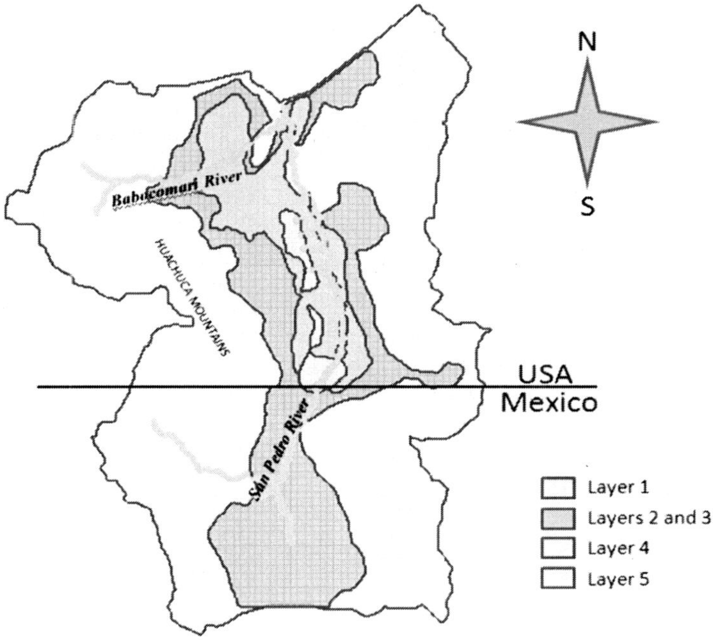

Figure 10. Model layers [2] in plan view with San Pedro River intersecting various layers. Light blue line represents river location but does not necessarily signify perennial flow.

Use of Model in Water Resources Assessments

Initial groundwater modeling efforts by Pool and Dickinson [3] simulated transient groundwater levels and baseflows in the basin resulting from 20th-century (1902–2003) pumping and changes in riparian evapotranspiration associated with climate and evolving geomorphology of the stream channel [14]. These simulations reflected the development of a major cone of depression under the Sierra Vista/ Fort Huachuca population center on the west side of the river, and a smaller area of groundwater depletion very close to the river near the communities of Palominas and Hereford, where agricultural pumping occurred for many decades (Figure 3a). The effects of groundwater use in Mexico also manifest as an irregular cone of depression progressing northward from the south edge of the regional aquifer.

Simulations by Lacher [33] built on the work of Pool and Dickinson [3] and projected pumping through the 21st century based on population projections developed by the Arizona Department of Commerce [54] and TischlerBise [55], and calculating the total drawdown (Figure 3) and change in baseflow over the 1902–2100 period (Figure 11). These simulations predicted widespread increases in aquifer storage depletion across the western side of the basin during the 21st century (Figure 3b,c). They also quantified projected declines in baseflow in the basin over the next 100 years due to pumping and evapotranspiration, as well projected increases in baseflow until about 2050 resulting from the recharge facility at the City of Sierra Vista's Environmental Operations Park (EOP) (Figure 11b).

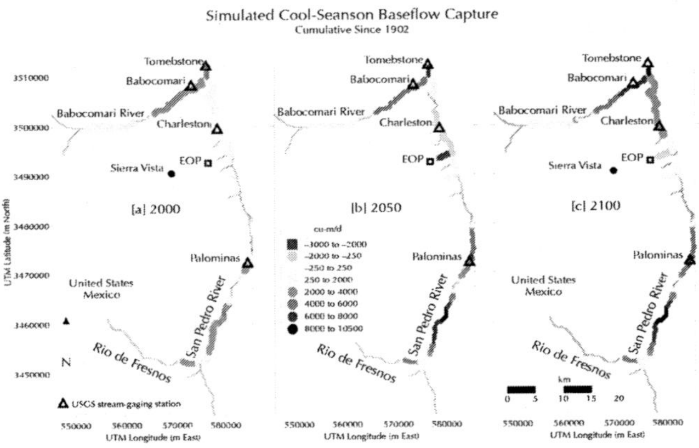

Figure 11. Simulated baseflow capture from 1902 to: (**a**) 2000; (**b**) 2050; and (**c**) 2100. Capture is cumulative [33] and measured in cu-m/d. EOP indicates location of Sierra Vista's Environmental Operations Park wastewater recharge facility.

Water budget calculations and groundwater simulations suggest that cumulative total groundwater depletion is presently on the order of 800 MCM in the subwatershed with annual net storage loss over the last decade on the order of 5 to 7 MCM [56,57]. Policy makers in the subwatershed have considered the feasibility of importing water at a maximum rate of roughly 37 MCM/yr [38]. However, even that most optimistic rate of importation would require nearly 25 years to bring the subwatershed's cumulative water budget back into balance. In the meantime, ongoing pumping-induced baseflow capture would continue to depress the groundwater gradient between the pumping centers and the river (Figure 4) further reducing baseflows.

Near-Stream Simulated Recharge Site Selection

Although artificial recharge of urban-enhanced runoff through detention basins has long been considered a viable option for mitigating impacts of groundwater pumping, the basins have previously been designed to target the active cone of depression in the Sierra Vista/Fort Huachuca area [34] rather than the river. The concept of simulating targeted near-stream recharge arose from the process of quantifying the pumping-induced capture of baseflow (*i.e.*, a reduction of groundwater discharge from the regional aquifer to the river) over time using the regional groundwater flow model [33]. With simulated current baseflow capture for the entire basin and projections of aquifer storage and baseflow capture trends over the next century, near-stream recharge was identified as a potential mechanism for addressing projected baseflow "deficits" (declines from 2003 (end of transient model calibration period) baseline values) in targeted areas of the basin.

Taken together, the suite of hydrologic tools pointed to sections of the San Pedro River where greater protection was warranted and where meaningful impacts—such as converting an intermittent stream reach into perennial reach—could be made. Simulations with the groundwater model [58] suggested that developing a distributed, strategically located network of recharge projects near select river reaches might result in groundwater mounding that would effectively compensate for baseflow capture, thus, protecting baseflow and the riparian vegetation community from the anticipated effects of pumping for several decades or more.

Application of Recharge Models

A small core team of technical experts within the Upper San Pedro Partnership Technical Committee selected three trial sites for simulating hypothetical recharge near the river: (1) Palominas; (2) Garden Canyon; and (3) Babocomari (Figure 12). The site selection process was informed by both simple geography (upper, middle, and lower part of the subwatershed) and current and projected baseflow capture in the river system. Simulated recharge at each trial site consisted of surface recharge over four 250 m × 250 m model cells (0.25 km^2) for the period 2012–2111. For each of the three trial sites shown in Figure 12, the recharge simulation investigation involved increasing recharge, as needed, to prevent any decline in baseflow below baseline (2003) levels downstream of the site over the simulation period while also preventing simulated surface flooding at the trial site.

Starting with 0.62 MCM/yr, simulated recharge at each site was incrementally increased, and baseflow response tested, until baseflow downstream of each site remained at or above 2003 levels for the 100-year simulation period. Figure 13 shows simulated change in the cool season (October–March) baseflow from 2003 (end of the transient calibration period) conditions in the years 2030, 2050, 2070, and 2111. As the decreasing cool (blue and green) colors in the northern half of the subwatershed over time indicate, baseflows are predicted to fall below 2003 levels in all of the mainstem San Pedro River and on the Babocomari by 2111. Recharge at the Sierra Vista EOP successfully maintains simulated baseflows in the mainstem above 2003 levels until at least 2070, but the impacts of pumping (deepening and widening cone of depression) overwhelm the recharge benefits by about the turn of the century.

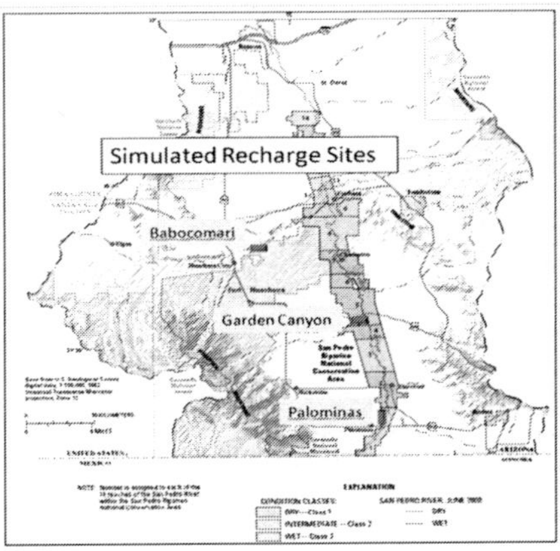

Figure 12. Simulated trial recharge sites (Babocomari, Garden Canyon, and Palominas) in the Sierra Vista subbasin. Riparian condition class reaches de-

lineated within the SPRNCA [10]. Adapted from Figure 42 in Stromberg and others [10].

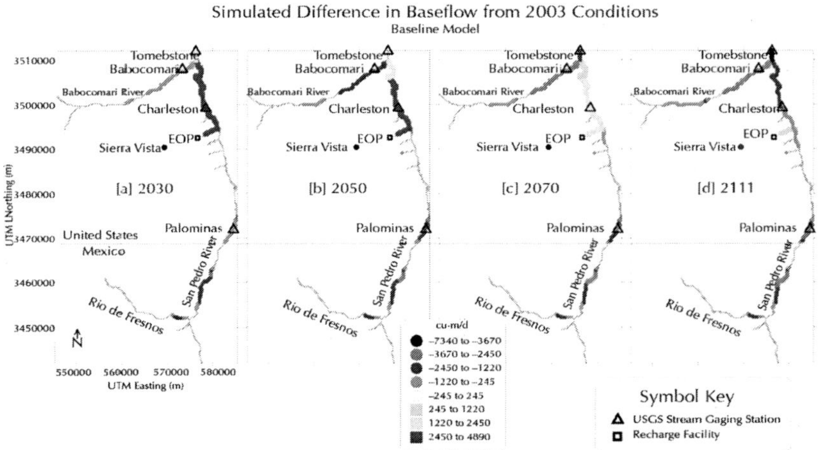

Figure 13. Simulated difference in baseflow (cu-m/d) in basin streams from 2003 conditions in: (a) 2030; (b) 2050; (c) 2070; and (d) 2111.

Simulation Results

Hydraulic conductivity, antecedent depth to groundwater, and projected aquifer storage depletion over time controlled the simulated baseflow response to recharge at each of the three trial sites. Since simulated recharge applied at each of the three trial sites was tailored to meet the anticipated baseflow deficit downstream, each site demanded a unique recharge distribution and exhibited a unique response to the simulated recharge. Figure 14 illustrates the recharge rates determined by trial and error as necessary to sustain simulated baseflows at or above 2003 levels in the groundwater model [3] through 2111. While each of the three test sites exhibited a unique response to the simulated recharge rates shown in Figure 14, the Babocomari site exhibited a much higher demand for recharge and a much more pronounced response in the underlying groundwater than the other two sites [58]. For the purpose of illustration, only the Babocomari test site recharge results will be discussed in detail here.

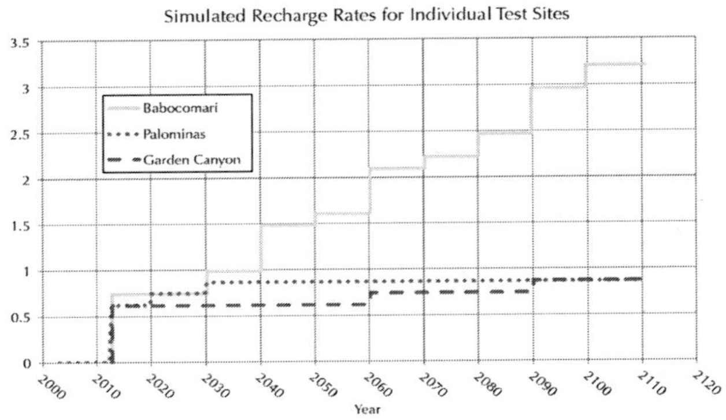

Figure 14. Simulated recharge rates for the Babocomari, Palominas, and Garden Canyon test sites.

Figure 15 illustrates the groundwater response to recharge at the Babocomari site. The black line represents depth to groundwater (DTW) under the stream adjacent to the recharge site in the baseline model with no simulated recharge at the Babocomari site. The orange line shows DTW in response to a constant rate of 0.62 MCM/yr. recharge, and the blue line shows DTW in the same location under the stream for the varying-recharge scenario illustrated by the green "Variable Recharge Rate" line. In the absence of any intervention (black DTW curve), simulated heads at this site are projected to drop by more than 10 m over the 21st century in response to pumping. However, these simulations suggest that incrementally increasing recharge from an initial rate of 0.74 MCM/yr in 2012 to 3.2 MCM/yr by 2100 would successfully maintain groundwater levels under the river at or slightly above the 2003 level.

The oscillation in the blue DTW curve reflects the fact that the simulated variable recharge increased groundwater levels under the river to a depth between the top of the evapotranspiration (ET) surface (1.5 m below top of aquifer) and the ET extinction depth at 6 m below the top of the aquifer. Thus, groundwater is more accessible to riparian vegetation in the varying-rate recharge scenario than in the baseline case, but baseflows still remain at or above 2003 levels in the area of the stream downstream where baseflow declines are projected under baseline conditions.

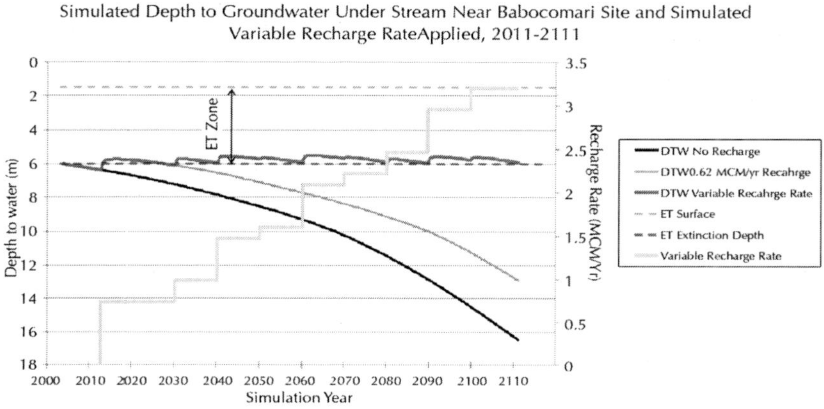

Figure 15. Simulated variable recharge rate at the Babocomari site and depth to water (DTW) under stream adjacent to the Babocomari recharge site from 2012 to 2111 for three scenarios: (a) no recharge; (b) 0.62 MCM constant-rate recharge; and (c) variable-rate recharge scenarios [58]. Evapotranspiration (ET) zone occurs between 1.5 and 6 m [3].

Optimization of Recharge Rates

Section 2.3.5 presented the results of simulated varying-rate recharge at each of three trial sites individually along the San Pedro and Babocomari Rivers. One additional simulation combining the three trial recharge sites was run in order to answer the question of whether hydrologic efficiencies might be gained with simultaneous recharge at all three sites [58]. Because the reach of the San Pedro affected by simulated recharge at the Garden Canyon site is downstream of both the Babocomari and Palominas sites, some reduction in the required recharge at the Garden Canyon site was achieved by combining all three trial recharge sites in a single simulation. Figure 16 illustrates the reduced recharge required at the Garden Canyon site to maintain simulated baseflows downstream of that site at 2003 levels when recharge is simulated at the Palominas and Babocomari trial sites concurrently.

Total recharge required to maintain simulated baseflows at 2003 levels downstream of each of the three test sites is shown in Figure

17. The blue curve shows the total recharge required when the three sites are simulated independently of each other, and the green curve shows the total recharge requirement when all three test sites are simulated concurrently. The difference between the curves illustrates the efficiency gained by combining the three recharge test sites. The maximum total recharge rate in the independent simulations is 4.93 MCM/yr (3.35 MCM/yr average over the 2012–2111 period), but that value drops to 4.63 MCM/yr (2.91 MCM/yr average) for the concurrent recharge simulations. The average recharge saved by operating all three sites concurrently is 0.45 MCM/yr. Figure 18 shows the final simulated change in baseflow from 2003 conditions when recharge at the three test sites is optimized for concurrent simulation of the three sites.

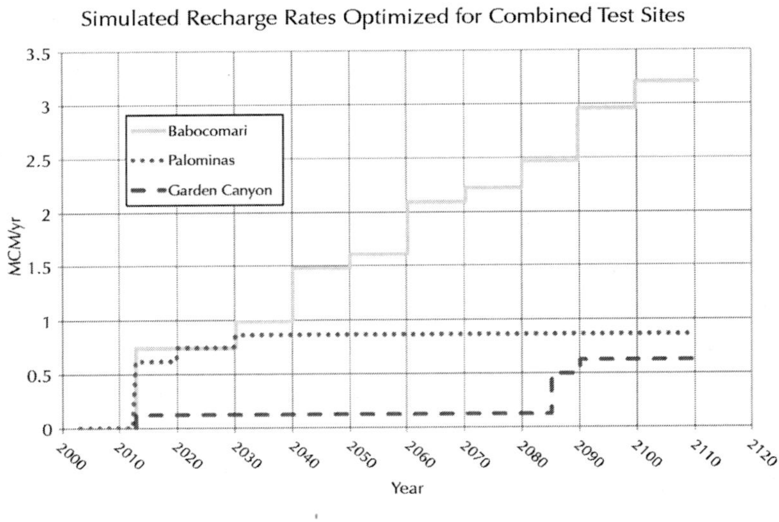

Figure 16. Simulated variable recharge rates for the Babocomari, Palominas and Garden Canyon test sites. Simulating recharge at all three sites concurrently allowed a reduction in Garden Canyon recharge relative to the rates for the independent recharge simulations.

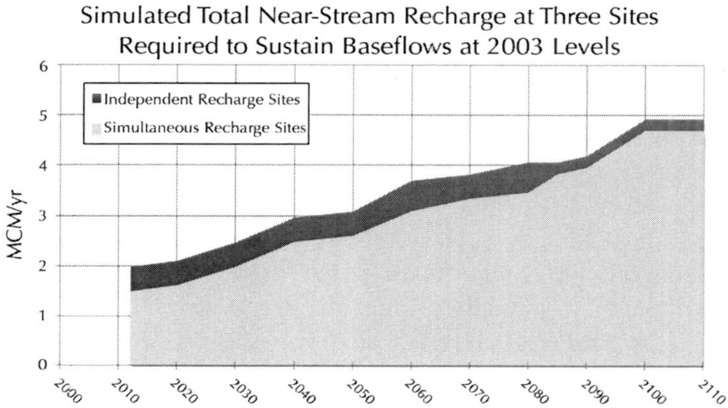

Figure 17. Total simulated recharge at the Babocomari, Palominas, and Garden Canyon test sites required to maintain baseflows downstream of each site at or above 2003 levels when recharge is simulated at each site independently and when all three sites are simulated concurrently.

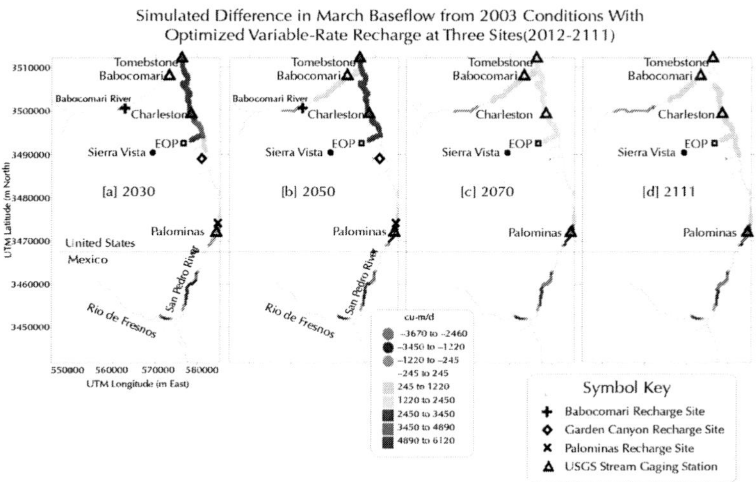

Figure 18. Simulated difference in March baseflow (cu-m/d) from 2003 conditions with optimized variable-rate recharge at three trial recharge sites in: (a) 2030; (b) 2050; (c) 2070; and (d) 2111.

Advantages/Disadvantages and Evolution of the Tool

Any perceived bias in the construction of the model, or simple disagreements among stakeholders with the technical modeling approach, can be problematic in terms of how modeling results will be used, if at all. Engagement of stakeholders throughout the model development process is essential for it to be embraced as a useful tool for decision making among varied interests. In this case, the groundwater model [3] was developed in response to wide dissatisfaction with some precursor models of the subwatershed. Richter and others [1] describe the process by which stakeholders were involved in the development of the Pool and Dickinson model [3] and the evolution of trust in the model among technical water resources experts and politicians in the basin.

Several years of experience with the Pool and Dickinson model [3] in developing and communicating results of baseline projections helped pave the way for using it as a tool to evaluate the prospects of near-stream recharge. The model made experimenting with various recharge rates and tracking the resultant changes in baseflow a fairly quick and inexpensive undertaking. While we feel that these simulation efforts successfully conveyed the general concept and potential merits of near-stream recharge to the public and decision makers, significant criticism arose from our choice of an arbitrary initial recharge rate of 0.62 MCM/yr for the hypothetical recharge sites. We viewed the initial near-stream recharge simulations as an exploratory mission to determine how much water would be required to produce the desired effect on baseflows in the San Pedro River, irrespective of the potential feasibility of attaining that quantity of water or distributing it in the locations of interest. As our initial simulation results were presented to stakeholders, many of them questioned the value in simulating recharge with water that is not available and for which no plans to develop were pending. We saw the near-stream modeling process as a "proof-of-concept" effort, but others quickly made the leap to the real difficulties in securing water for recharge. Significant time and energy were expended in efforts to bridge this conceptual gap, and in hindsight, more early effort to clarify the purpose and strategy behind the simulations would have been helpful.

Application of Recharge Simulations to Upper San Pedro Parcels

The primary value in the groundwater modeling efforts undertaken for the three "test sites" described above was in proving the potential benefit of multiple recharge sites near the river. In recent years, the U.S. Army Compatible Use Buffer program, Cochise County, and The Nature Conservancy collaboratively acquired and set aside for conservation purposes four parcels on the west side of the San Pedro River totaling 2226 hectares within the subwatershed (Figure 5). Collectively, these properties make up the physical sites currently under consideration for development of a network of near-stream recharge projects. The groundwater modeling process detailed above for the three hypothetical "test sites" was employed to some extent in the preliminary planning process for parcels 1 and 3 in Figure 5. As projects move from the conceptual phase to the physical site investigation stage, use of the groundwater model is being adapted to suit the project planning needs for the individual sites. While development of a recharge project on parcel 1 was constrained by several factors (parcel size, flood-control objectives, location of high-permeability soils off site), the current planning process for parcel 3 is relatively unconstrained. Modifications to model structure to reflect observed field conditions and refinement of the model grid to allow for more detailed simulation of potential recharge are two of the anticipated outcomes of the next phase of investigation at parcel 3.

RESULTS AND DISCUSSION

The suite of analytical tools discussed here is being used to inform key decisions necessary to balance groundwater use and maintain San Pedro River flows and associated riparian area ecological health. An extensive collection of hydrological studies and a robust, long-term monitoring program in the San Pedro basin have provided policy makers and stakeholders with important information about the complex relationships between groundwater condition, streamflow, and the ecological integrity of the riparian system within and near the SPRNCA. As a result, most of the stakeholders in the subwatershed understand that much of the San Pedro's riparian vegetation uses groundwater from the stream alluvium, and that this alluvial aquifer stores water from flood flows, receives groundwater from the regional aquifer, and contributes baseflow to the river during low-flow periods.

The groundwater model used in this study does not incorporate the complex interactions between flood-driven recharge of the shallow alluvium which influences baseflow to varying degrees from year to year, and pumping-induced depletions of the regional aquifer, which take decades to centuries to alter baseflow. While the model's authors made every effort to exclude storm-flow influences from the baseflow measurements they used in model calibration [3], it is virtually impossible to ensure that any given measured baseflow value represents only regional aquifer groundwater. The implication of this limitation is that the model may slightly overestimate the regional aquifer's contribution to baseflow, which would manifest as overestimated aquifer transmissivity and/or streambed conductance. That outcome, in turn, means that the model would also overestimate the simulated impact of pumping on baseflow until capture reaches its maximum (equilibrium) value. For the purpose of this study, which is to develop strategies for mitigating pumping-related impacts on baseflow, that response would be conservative and acceptable.

The Pool and Dickinson [3] model represents the culmination of years of study and is the best tool currently available for the study area. As discussed by Richter and others [1], using a groundwater model that is accepted by the vast majority of decision makers to perform predictive groundwater modeling has been essential for beginning to make management and project decisions from a common starting point. The modeling has demonstrated that within the next 100 years, two regional cones of depression will enlarge and likely change the nature of the hydrologic connection between the San Pedro River and the regional aquifer, reducing baseflow and impacting the dependent riparian system and thus wildlife populations.

Awareness and acceptance of an impending problem, however, is only the first step in finding a solution. The additional tools of the wet/dry maps and groundwater capture/recharge maps helped to focus management attention on finding both the most vulnerable areas of the system and the most beneficial locations for mitigation efforts. Analysis of wet/dry maps showing surface water presence during the driest time of year and areas with high year-to-year variation in wetted length may be the first physical evidence of changes in local groundwater conditions at the river. While wetted length is not solely controlled by groundwater conditions (it may also be affected by climate), different trends in various reaches of the river may help identify areas at higher risk of future ecological changes. Aligning these low-flow river reaches with the groundwater capture maps provided a rough indication of the rate that recharge in those areas of the aquifer might respond, expressing itself as baseflow in the San Pedro River. The capture maps also suggested the suitability of various locations for recharge that may communicate with the alluvial aquifer and/or the San Pedro River. On-site investigation of actual hydrogeologic conditions and suitability for recharge at a particular location is the first step toward refining a preliminary conceptual model derived from the tools described in this paper.

In the San Pedro basin, the use of these complementary approaches informed the purchase of the most hydrologically sensitive lands near or adjacent to SPRNCA (Figure 5) in order to both defer residential and/or agricultural development and provide the opportunity for near-stream recharge project development. The concept of a strategically located network of recharge projects near these river reaches evolved, in part, from the success of more than a decade of managed aquifer recharge at the City of Sierra Vista Environmental Operations Park in supporting both baseflow and replenishing the deep regional aquifer. We anticipate that recharging urban-enhanced runoff, storm water, and treated effluent near at-risk reaches shown would create groundwater mounds to sustain surface water flow and supplement alluvial groundwater levels during low-flow periods, effectively compensating for the otherwise deleterious impacts of encroaching cones of depression in the regional aquifer on baseflows for the next several decades or more.

CONCLUSIONS

The application of the study methods presented in this paper and the development of system-specific techniques appear to have great promise for protecting dry-land riparian systems from the impacts of groundwater extraction, surface water diversions, and the extremes of climate change for up to several decades. The ecological, cultural, and economic significance of the San Pedro River has made it one of the most well-studied and understood river systems in the world. The tremendous volume of data and hydrologic tools already developed for this specific system coupled with many years of collaborative partnerships that have matured with the science make the San Pedro basin a very unique policy environment. There is likely no other river system with an identical set of social, political, and ecological circumstances, but the hydrologic analysis tools described in this paper can be used anywhere. The strength of collaborative partnerships and knowledge of which tools they jointly support is key to building a common understanding of the history, goals, and resources at hand to make real progress.

ACKNOWLEDGMENTS

The authors would like to acknowledge the Walton Family Foundation for its support in funding much of the science and policy development in the Upper San Pedro Basin in recent years. They would also like to express gratitude to the reviewers whose collective input substantially improved this paper.

REFERENCES

1. Richter, H.E.; Gungle, B.; Turner, D.S.; Bushman, B.; Lacher, L.J. Development of a Shared Vision Toward Sustainability of Groundwater to Protect and Sustain Baseflows of the Upper San Pedro River, Arizona. *Water* 2014, *6*, 2519–2538.
2. United States Congress. Arizona-Idaho Conservation Act of 1988. *Public Law 100-696*, Title I. 1988.
3. Pool, D.R.; Dickinson, J.E. *Ground-Water Flow Model of the Sierra Vista Subwatershed and Sonoran Portions of the Upper San Pedro Basin, Southeastern Arizona, United States, and Northern Sonora, Mexico*; U.S. Geological Survey: Reston, VA, USA, 2007; p. 47.
4. Cooperative Climatological Data Summaries for Arizona. Data for Coronado National Monument, Sierra Vista, and Tombstone

stations. Available online: http://www.wrcc.dri.edu/summary/Climsmaz.html (accessed on 13 June 2014).

5. Thomas, B.E.; Pool, D.R. *Trends in Streamflow of the San Pedro River, Southeastern Arizona, and Regional Trends in Precipitation and Streamflow in Southeastern Arizona and Southwestern New Mexico*; U.S. Geological Survey Professional Paper 1712. U.S. Geological Survey: Reston, VA, USA, 2007; p. 79.
6. 2013–2050 Sub-county Population Projections. Available online: http://www.workforce.az.gov/population-projections.aspx (accessed on 13 June 2014).
7. July 1, 2013 Population estimates for Arizona's counties, incorporated places and unincorporated balance of counties. Available online: http://www.workforce.az.gov/pubs/demography/July%201_2013%20PopulationEstimates_Final.pdf (accessed on 13 June 2014).
8. Carreira, R. Economic Impact of Fort Huachuca. Undated white paper. Center for Economic Research at Cochise College. Available online: http://www.cochise.edu/cfiles/files/CER/CER%20Studies/Economic_Impact_of_Fort_Huachuca.pdf (accessed on 13 June 2014).
9. Skagen, S.K.; Melcher, C.P.; Howe, W.H.; Knopf, F.L. Comparative use of riparian corridors and oases by migrating birds in southeast Arizona. *Conserv. Biol.* 1998, *12*, 896–909.
10. Stromberg, J.C.; Lite, S.J.; Dixon, M.; Rychener, T.; Makings, E. Relations between streamflow regime and riparian vegetation composition, structure, and diversity within the San Pedro Riparian National Conservation Area, Arizona. In *Hydrologic Requirements of and Consumptive Ground-Water Use by Riparian Vegetation along the San Pedro River, Arizona*; Leenhouts, J.M., Stromberg, J.C., Scott, R.L., Eds.; U.S. Geological Survey Scientific Investigations Report 2005-5163. U.S. Geological Survey: Reston, VA, USA, 2006; pp. 57–106.
11. Stromberg, J.C.; Bagsted, K.J.; Leenhouts, J.M.; Lite, S.J.; Making, E. Effects of stream flow intermittency on riparian vegetation of a semiarid region river (San Pedro River, Arizona). *River Res. Appl.* 2005, *21*, 925–938.
12. Haynes, C.V., Jr. Curry Draw, Cochise County, Arizona: A late Quaternary stratigraphic record of Pleistocene extinction and

paleo-Indian activities. In *Geologic Diversity of Arizona and Its Margins: Excursions to Choice Areas*; Arizona Bureau of Geology and Mineral Technology Special Paper 5. Davis, G.H., VandenHolder, E.M., Eds.; U.S. Department of the Interior: Phoenix, AZ, USA, 1987; pp. 23–28.

13. U.S. Bureau of Land Management. Tucson Field Office Paleontology Sites. Available online: http://www.blm.gov/az/st/en/prog/cultural.html (accessed on 15 June 2014).
14. Hereford, R. *Entrenchment and Widening of the Upper San Pedro River, Arizona*; Special Paper 282. Geological Society of America: Boulder, CO, USA, 1993; p. 46.
15. Arizona Department of Water Resources. Surface Water Adjudications. Available online: http://www.azwater.gov/AzDWR/SurfaceWater/Adjudications/ (accessed on 15 June 2014).
16. U.S. Department of Agriculture—Agricultural Research Service Experimental Watersheds. Available online: http://www.ars.usda.gov/Research/docs.htm?docid=10978&page=2 (accessed on 15 June 2014).
17. Brown, S.G.; Davidson, E.S.; Kister, L.R.; Thomsen, B.W. *Water Resources of Fort Huachuca Military Reservation, Southeastern Arizona*; U.S. Geological Survey Water-Supply Paper 1819–D. U.S. Geological Survey: Reston, VA, USA, 1966; pp. D1–D57.
18. Brown, S.G.; Aldridge, B.N. *Streamflow Gains and Losses and Ground-Water Recharge in the San Pedro River Basin, Arizona*; USGS Administrative Report. U.S. Geological Survey: Reston, VA, USA, 1973; p. 45.
19. Freethey, G.W. *Hydrologic Analysis of the Upper San Pedro Basin from the Mexico–United States International Boundary to Fairbank, Arizona*; U.S. Geological Survey Open-File Report 82–752. U.S. Geological Survey: Reston, VA, USA, 1982; p. 52.
20. Vionnet, L.B.; Maddock, T. *Modeling of Ground-Water Flow and Surface Water/Groundwater Interactions in the San Pedro River Basin—Part I—Cananea, Mexico to Fairbank, Arizona*; HWR Report No. 92-010. University of Arizona: Tucson, AZ, USA, 1992; p. 96.

21. Corell, S.W.; Putman, F.; Lovvik, D.; Corkhill, F. *A Groundwater Flow Model of the Upper San Pedro Basin, Southeastern Arizona*; Phoenix, Arizona Department of Water Resources Modeling Report No. 10. Arizona Department of Water Resources: Phoenix, AZ, USA, 1996; p. 85.
22. Goode, T.; Maddock, T., III. *Simulation of Groundwater Conditions in the Upper San Pedro Basin for the Evaluation of Alternative Futures*; HWR report No. 00-30. Department of Hydrology and Water Resources, University of Arizona: Tucson, AZ, USA, 2000.
23. Arizona Department of Water Resources. *Hydrographic Survey Report (HSR) for the San Pedro River Watershed*; Arizona Department of Water Resources: Phoenix, AZ, USA, 1990; p. 548.
24. Arizona Department of Water Resources. *Upper San Pedro Basin Active Management Area Report, March 2005*; Arizona Department of Water Resources: Phoenix, AZ, USA, 2005.
25. Pool, D.R.; Coes, A.L. *Hydrogeologic Investigations of the Sierra Vista Sub-Watershed of the Upper San Pedro Basin, Cochise County, Southeast Arizona*; USGS Water Resources Investigations Report 99-4197. U.S. Geological Survey: Tucson, AZ, USA, 1999; p. 41.
26. Cooperative water use management related to Fort Huachuca, Arizona, and Sierra Vista subwatershed. *U.S. Public Law 108-136*, Section 321. 2003.
27. Upper San Pedro Partnership. *Water Management of the Regional Aquifer in the Sierra Vista Subwatershed, Arizona—2011 Report to Congress*; U.S. Department of Interior: Washington, DC, USA, 2013; p. 16.
28. Coes, A.L.; Pool, D.R. *Ephemeral-Channel and Basin-Floor Infiltration in the Sierra Vista Subwatershed, Arizona*; U.S. Geological Survey Open-File Report 05–1023. U.S. Geological Survey: Reston, VA, USA, 2005; p. 67.
29. Gungle, B. *Timing and Duration of Flow in Ephemeral Streams of the Sierra Vista Subwatershed of the Upper San Pedro Basin, Cochise County, Southeastern Arizona*; U.S. Geological Survey Scientific Investigations Report 2005–5190. U.S. Geological Survey: Reston, VA, USA, 2006; p. 47.

30. *Hydrologic Requirements of and Consumptive Ground-Water Use by Riparian Vegetation Along the San Pedro River, Arizona*; Leenhouts, J.M., Stromberg, J.C., Scott, R.L., Eds.; U.S. Geological Survey Scientific Investigations Report 2005-5163. U.S. Geological Survey: Reston, VA, USA, 2006; p. 154.
31. Leake, S.A.; Pool, D.R.; Leenhouts, J.M. *Simulated Effects of Ground-Water Withdrawals and Artificial Recharge on Discharge to Streams, Springs, and Riparian Vegetation in the Sierra Vista Subwatershed of the Upper San Pedro Basin, Southeastern Arizona*; U.S. Geological Survey Scientific Investigations Report 2008-5207. U.S. Geological Survey: Reston, VA, USA, 2008; p. 14.
32. Kennedy, J.R.; Gungle, B. *Quantity and Sources of Baseflow in the San Pedro River near Tombstone, Arizona*; U.S. Geological Survey Scientific Investigations Report 2010–5200. U.S. Geological Survey: Reston, VA, USA, 2010; p. 43.
33. Lacher, L.J. *Simulated Groundwater and Surface Water Conditions in the Upper San Pedro Basin, 1902–2105*; Lacher hydrological Consulting: Tucson, AZ, USA, 2011.
34. Stantec. *Cochise County Flood Control/Urban Runoff Recharge Plan*; Stantec Consulting, Inc.: Tucson, AZ, USA, 2006.
35. GeoSystems Analysis, Inc. *Hydrologic and Incidental Groundwater Recharge Study*; Stantec Consulting, Inc.: Tucson, AZ, USA, 2006; p. 70.
36. U.S. Department of Agriculture—Agricultural Research Station. Automated Geospatial Watershed Assessment Tool, developed in cooperation with the U.S. Environmental Protection Agency. Available online: http://www.tucson.ars.ag.gov/agwa/index.php?option=com_content&view=article&id=5&Itemid=26 (accessed on 14 June 2014).
37. GeoSystems Analysis, Inc. *Storm Water Recharge Feasibility Analysis, Appendix B: AGWA/KINEROS Simulations for Coyote Wash Watershed*; Report to Cochise County for Project SP-0011. GeoSystems Analysis, Inc.: Tucson, AZ, USA, 2004.
38. United States Bureau of Reclamation-Lower Colorado Region. *Appraisal Report: Augmentation Alternatives for the*

Sierra Vista Subwatershed, Arizona; U.S. Department of Interior: Washington, DC, USA, 2007; p. 77.

39. United States Bureau of Reclamation. Grant No. 00-FC-32-0030 Between Bureau of Reclamation Phoenix Area Office and The City of Sierra Vista Pertaining to Construction of the Sierra Vista Water Reclamation Facility Effluent Recharge Demonstration Project. February 2000.
40. Brown and Caldwell. *Sierra Vista Water Reclamation Facility Groundwater Modeling Study*; Prepared for City of Sierra Vista. Brown and Caldwell: Sierra Vista, AZ, USA, 2009.
41. Turner, D.; Richter, H. Wet/Dry Mapping: Using Citizen Scientists to Monitor the Extent of Perennial Surface Flow in Dryland Regions. *Environ. Manag.* 2011, *47*, 497–505.
42. Theis, C.V. The source of water derived from wells: Essential factors controlling the response of an aquifer to development. *Civ. Eng.* 1940, *10*, 277–260.
43. Lohman, S.W.; Bennett, R.R.; Brown, R.H.; Cooper, H.H., Jr.; Drescher, W.J.; Ferris, J.G.; Johnson, A.I.; McGuiness, C.L.; Piper, A.M.; Rorabaugh, M.I.; *et al. Definitions of Selected Ground-Water Terms—Revisions and Conceptual Refinements*; U.S. Geological Survey Water-Supply Paper 1988. U.S. Geological Survey: Reston, VA, USA, 1972; p. 21.
44. Bredehoeft, J.D.; Papadopulos, S.S.; Cooper, H.H. Groundwater: The water budget myth. In *Scientific Basis of Water-Resource Management, Studies in Geophysics*; National Academy Press: Washington, DC, USA, 1981; pp. 51–57.
45. Brown, R.E. *The Cone of Depression and the Area of Diversion around a Discharging Well in an Infinite Strip Aquifer Subject to Uniform Recharge*; U.S. Geological Survey Water-Supply Paper 154SC. U.S. Geological Survey: Reston, VA, USA, 1963; pp. C69–C85.
46. Winter, T.C.; Harvey, J.W.; Franke, O.L.; Alley, W.M. *Ground Water and Surface Water: A Single Resource*; U.S. Geological Survey Circular 1139. U.S. Geological Survey: Reston, VA, USA, 1998; p. 15.

47. Heath, R.C. *Basic Groundwater Hydrology*; U.S. Geological Survey Water Supply Paper 2220. U.S. Geological Survey: Reston, VA, USA, 1986; p. 86.
48. Lohman, S.W. *Ground-Water Hydraulics*; U.S. Geological Survey Professional Paper 708. U.S. Geological Survey: Reston, VA, USA, 1972; p. 70.
49. Leake, S.A.; Reeves, H.W.; Dickinson, J.E. A new capture fraction method to map how pumpage affects surface water flow. *Groundwater* 2010, *48*, 690–700.
50. Barlow, P.M.; Leake, S.A. *Streamflow Depletion by Wells—Understanding and Managing the Effects of Groundwater Pumping on Streamflow*; U.S. Geological Survey Circular: 1376. U.S. Geological Survey: Reston, VA, USA, 2012; p. 95.
51. Leake, S.A.; Greer, W.; Watt, D.; Weghorst, P. *Use of Superposition Models to Simulate Possible Depletion of Colorado River Water by Ground-Water Withdrawal*; U.S. Geological Survey Scientific Investigations Report 2008-5189. U.S. Geological Survey: Reston, VA, USA, 2008; p. 25.
52. Harbaugh, A.W.; Banta, E.R.; Hill, M.C.; McDonald, M.G. *MODFLOW-2000, the U.S. Geological Survey Modular Ground-Water Model—User Guide to Modularization Concepts and the Ground-Water Flow Process*; U.S. Geological Survey Open-File Report 00-92. U.S. Geological Survey: Reston, VA, USA, 2000; p. 121.
53. Prudic, D.E. *Documentation of a Computer Program to Simulate Stream-Aquifer Relations Using a Modular, Finite-Difference, Ground-Water Flow Model*; U.S. Geological Survey Open-File Report 88-729. U.S. Geological Survey: Reston, VA, USA, 1989; p. 113.
54. Arizona Department of Commerce (AzDC) population projections for 2006, 2009, and 2010. Available online: http://www.azcommerce.com/EconInfo/Demographics/Population+Projections.htm (accessed on 30 October 2010).
55. TischlerBise. *Infrastructure Improvements Plan and Development Fee Study*; consulting report prepared for the City of Sierra Vista. TischlerBise: Bethesda, MD, USA, 2009.

56. *Upper San Pedro Partnership—Water Management of the Regional Aquifer in the Sierra Vista Subwatershed, Arizona—2004 Report to Congress*; U.S. Department of Interior: Washington, DC, USA, 2005; p. 36.
57. *Upper San Pedro Partnership—Water Management of the Regional Aquifer in the Sierra Vista Subwatershed, Arizona—2010 Report to Congress*; U.S. Department of Interior: Washington, DC, USA, 2011; p. 20.
58. Lacher, L.J. *Simulated Near-Stream Recharge at Three Sites in the Sierra Vista subbasin, Arizona*; Lacher hydrological Consulting: Tucson, AZ, USA, 2012.

Chapter 6

Economics of Managed Aquifer Recharge

Robert G. Maliva

Schlumberger Water Services, 1567 Hayley Lane, Suite 202, Fort Myers, FL 33907, USA

ABSTRACT

Managed aquifer recharge (MAR) technologies can provide a variety of water resources management benefits by increasing the volume of stored water and improving water quality through natural aquifer treatment processes. Implementation of MAR is often hampered by the absence of a clear economic case for the investment to construct and operate the systems. Economic feasibility can be evaluated using cost benefit analysis (CBA), with the challenge of monetizing benefits. The value of water stored or treated by MAR systems can be evaluated by direct and indirect measures of willingness to pay including market price, alternative cost, value marginal product, damage cost avoided, and contingent value methods. CBAs need to incorporate potential risks and uncertainties, such as failure to meet performance objectives. MAR projects involving high value uses, such as potable supply, tend to be economically feasible provided that local hydrogeologic conditions are favorable. They need to have low construction and operational costs for lesser value uses, such as some irrigation. Such systems should therefore be financed by project beneficiaries, but dichotomies may exist between beneficiaries and payers. Hence, MAR projects in developing countries may be economically viable, but external support is often required because of limited local financial resources.

INTRODUCTION

Managed aquifer recharge (MAR) is defined as "intentional banking and treatment of waters in aquifers" [1]. The term MAR was introduced as an alternative to "artificial recharge", which has the connotation that the use of the water was in some way unnatural [1]. MAR includes a great diversity of technologies to store and treat water including aquifer storage and recovery (ASR), infiltration basins, salinity barriers, soil-aquifer treatment, and riverbank filtration. The water resources management benefits of MAR are compelling. However, the question arises as to why MAR has not yet been implemented to an even greater degree. The answer often lies in that decision makers, such as water utility managers, water management agency officials, and political leaders, have not been provided an equally compelling, sound economic case for investment in the technologies.

Investments in infrastructure, whether for water or other purposes, need to be justified in terms of the benefits of the project equaling or exceeding the construction and operational costs. The costs of the projects should also be less than the costs of alternative projects that provide the same benefits. Cost–benefit analysis (CBA) can be used to evaluate MAR projects, where their costs and benefits can be accurately quantified in monetary terms. However, economic analyses of water projects are often hampered by the difficulty of accurately quantifying the value of water, which can vary greatly depending upon circumstances. Todd [2] in a pioneering paper noted with respect to the economics of groundwater recharge that "It is clear the analysis of the benefits of artificial recharging is dependent on what value can be assigned to a unit volume of water" and "in assessing the benefits of artificial recharge, consideration must be given to the importance of water to the total economy, to the value of water for various uses, as well as to the direct and intangible benefits that may accrue."

Water and wastewater projects are not necessarily evaluated solely based on their profitability to the system owner and operator. Water and wastewater utilities often have mandates to provide specified levels of services irrespective of the profitability of each individual system component. Water has social and environmental values as a necessity of life. Hence, water is often provided to poor communities even if the revenues generated do not cover costs. Governmental projects are also often funded all or in part by general revenues (rather than entirely from revenues from the sale of water) with the goal of achieving societal benefits. MAR projects may thus be economically evaluated by comparison to non-managed scenarios [3] or other water management or treatment options to achieve the same goals [4, 5]. The feasibility of MAR projects also depends upon financial feasibility [6], which addresses whether funding is available for a project and how a project will be paid for. In developing countries, MAR projects are available that may be economically feasible (*i.e.*, their benefits exceed costs) and could materially improve the quality of lives of the people, but financial resources are not available. Water projects often have to compete for limited financial resources against other types of projects (e.g., health, transportation) that also provide societal benefits.

The procedures for analyzing the economic benefits of groundwater presented by Bergstrom *et al.* [7] and the National Research Council [8] provide a basic framework for evaluating the economics of MAR systems. The first step in the evaluation is an analysis of the changes in groundwater quality and quantity resulting from the implementation of a MAR project. The change is evaluated relative to a reference state, which would normally be current conditions. The changes in groundwater services resulting from the change in groundwater quantity and quality is next evaluated. Finally, the economic value of the change in groundwater services is evaluated.

Water has been recognized to be an economic good, but its price is seldom set by a free market. Water also has social and environmental values that are difficult to quantify in monetary terms. Indeed, some people object to the very notion that economics should enter into decisions concerning water supply. The value of water also varies greatly depending on local circumstances. As water is critical for life, water can be priceless during extreme shortages. In some water-scarce developing countries, there are often large social costs associated with both physically obtaining the daily water supply and health impacts associated with poor water quality. On the contrary, during periods of abundant supply, the market value of water can be very low, and in the case of flooding it is a liability (*i.e.*, has a negative net value).

MAR SYSTEM TYPES AND BENEFITS

MAR includes a wide variety of processes by which water is intentionally added into an aquifer or induced to flow into and through an aquifer for treatment purposes. MAR, as defined by Dillon [1], includes two main end-member types of technologies: (1) methods that are used primarily to increase the volume of water stored in aquifers; and (2) methods that are used primarily for water or wastewater treatment. MAR systems with a water storage goal include ASR, aquifer recharge using wells and infiltration basins, and river channel modifications to enhanced aquifer recharge (e.g., check dams). MAR using wells, including specifically aquifer storage and recovery (ASR), was reviewed by Huisman and Olsthoorn [9], Pyne [10], and Maliva and Missimer [11]. Surface spreading methods were reviewed by Huisman and Olsthoorn [9], Oaksford [12] and Roscoe Moss Company [13]. The benefit of storage-type systems is the net increase in the volume of water stored in the aquifer. The increased storage results in an increase in the volume of water available for later beneficial use (abstraction benefits). Additional potential benefits result from the water being in place in the aquifer (*in-situ* benefits). *In-situ* benefits include reduced groundwater pumping costs, and avoidance of the need to replace or deepen production wells, restoration or maintenance of environmental (e.g., spring) flows, avoidance of land subsidence, and prevention of saline-water intrusion [2, 8].

MAR systems with a storage goal are primarily constructed in hydrological and engineering settings where there are at least periodic shortages of water and times when excess water is available that could be used to recharge aquifers. MAR is used in arid and semiarid lands, for example, to capture surface water that is episodically available during uncommon rainfall events. MAR is also employed in areas with humid climates, such as South Florida and parts of India, where there is a pronounced seasonality in rainfall. The systems are usually installed either where excess water is available (e.g., in-channel and off-channel infiltration systems in ephemeral streams and ASR systems at water treatment facilities) or where the water is used.

MAR systems with a primary treatment goal have been termed "aquifer recharge and recovery" (ARR) and include soil-aquifer-treatment (SAT) and aquifer storage transfer and recovery (ASTR). SAT is a high-rate land application system that involves the spreading of partially-treated domestic wastewater on the soil surface to provide natural treatment as the water infiltrates into the soil and flows through underlying aquifers. The vadose (unsaturated) zone is used as a natural filter to remove or reduce the concentrations of suspended solids, biodegradable organic matter, nutrients, metals, and pathogenic microorganisms, by a variety of filtration, sorption and biologically mediated reactions [14,15,16,17,18]. Additional filtration and removal of contaminants occurs as the water travels through the aquifer. ASTR involves the injection of water into an aquifer using wells and its recovery with separate production wells as a means of improving stored water quality by providing additional residence time and to take advantage of the filtration and other treatment processes provided by the aquifer [19]. The essential, defining feature of ASTR is the intentional use of water flow through an aquifer as a treatment method.

MAR systems vary greatly in their scale and thus how they should be evaluated by CBA. Large-scale systems owned and operated by water utilities or water management districts or agencies usually have well-defined costs and benefits, although there may be uncertainty in the quantification of benefits (e.g., monetary value of stored water in the absence of a free market). MAR also includes small-scale systems implemented in developing countries whose benefits, such as a reduction in labor, disease, and mortality due to the availability of a more convenient, reliable, and safer water supply, are difficult to express in monetary terms, but nonetheless have great value.

COST-BENEFIT ANALYSIS

According to the basic rule of benefit maximization, in which increasing the total value of scarce resources is assumed to be desirable, actions (such as the construction of MAR systems) should be undertaken if their total benefits exceed total costs [20]. Cost–benefit analysis (CBA) is addressed in microeconomic textbooks and some dedicated books (e.g., [21, 22]). Environmental CBA is a specific area of investigation (e.g., [23, 24, 25, 26]), which includes issues of water quality and supply.

The underlying goal of CBA is allocative efficiency. Policies should be adopted or investments made only if they provide net positive benefits. The policy or investment that yields the greatest net benefits should be selected. A limitation of CBA is that goals other than economic efficiency (e.g., equity and national security) may be of relevance to the policy [22]. CBAs are not performed in a moral vacuum and the social desirability of a particular set of costs and benefits may be a consideration [25]. However, even if decisions are not made solely on the basis of CBA, decisions should at least be informed by CBA such that it is at least an input into the decision-making process [25].

CBAs are commonly performed using the net present value (NPV) method, which considers both the initial investment in the project and benefits and costs expected to be achieved or incurred over the life of the project. Future benefits and costs are discounted at an appropriate rate. The basic NPV equation is

$$NPV = -C_0 + \sum B_i/(1 + r)^i - \sum C_i/(1 + r)^i \quad (1)$$

where C_0 is the initial (capital) costs in year 0; B_i and C_i are the benefits and costs in year "I" and "r" is the discount rate.

Cost-effectiveness analysis (least cost analysis) and lifecycle costs analysis consider only the costs to achieve a pre-set objective or criterion. Different options are considered that provide the same benefit or set of benefits. Cost-effectiveness analysis is suitable where valid and reliable estimation of benefits is not feasible [27]. It may be used to evaluate options to achieve a well-defined water supply or environmental goal. For example, if the decision is made to supply a given amount of potable water to a community as a social objective, then cost-effectiveness analysis could be used to evaluate different supply options. A limitation of cost-effectiveness analysis is that an entire list of projects could be ranked without any assurance that any of them are actually worth doing [25].

A basic requirement of CBAs is that accurate costs and benefits values be used. However, "appraisal optimism" is common, which is the tendency to exaggerate benefits and under-estimate cost escalations [25]. Appraisal optimism can be either accidental or intentional. In the latter case, those with a vested interest in a project may under-estimate costs or over-estimate benefits to gain support for a project, knowing that projects develop momentum for their continuation and thus become difficult to later terminate. For example, false economic analyses were widely used to give the perception that major water supply projects in the western United States made economic sense, when in fact they could never be economically justified because the farmers (the primary beneficiaries) could never afford the true cost of the delivered water [28].

The discount rate reflects time preference for benefits and costs, which varies between individuals. Individuals typically value a benefit more today, than they would value receiving the same benefit ten years from now. Discounting enables comparison of costs and benefits that occur at different times.

In economics, the discount rate is equal to the interest rate in a perfect capital market with no taxes or inflation [29]. Application of a discount factor reduces the importance of future costs and inevitably means that what happens long distances into the future has very little impact on decisions made today [30]. Discounting has been referred to as a "tyranny" that militates against the interests of future generations [24] and thus appears to be inconsistent with rhetoric and spirit of "sustainable development" as it violates the notion of intergenerational equity [25]. However, not discounting (*i.e.*, use of discount rate of zero) creates other problems in that the needs of generations very far into future are given equal weighting, which would encourage excessive saving at the expense of current needs [25]. Pearce *et al.* [25] present the arguments that a time-decreasing discount rate may be the most appropriate solution.

There is considerable disagreement as to what discount rate is appropriate. Freeman [29] suggested that a rate of 1%–4% is usually appropriate. Where the costs precede benefits, as is the case for most water projects, those who favor such projects may argue for a low rate while those who oppose them may argue for a high rate [23].

Not all costs and benefits of a MAR project are borne and accrued by the system owner. For example, all groundwater users in a basin may benefit from increased water levels in an aquifer resulting from a recharge program, whether or not they personally financially contribute to the project. Similarly, where a project receives external funding, such as a governmental grant, the system owner and participants may receive most or all the benefits of a system, while not having to pay the full costs. The results of a CBA that include all costs and benefits may thus differ from the results of a CBA that is limited only to the costs and benefits to the system owner. This dichotomy is addressed under finance.

It is important to also distinguish between financial CBA, which measures only the direct financial implications of a project, and social cost-benefit analysis, which measures the overall welfare impact of a project [31]. Social benefits associated with water projects include benefits associated with having a reliable, convenient, and safe source of water. Welfare impacts can also be considered to include environmental benefits and costs. Valuation of welfare effects in monetary terms brings with it problems and can lead to inappropriate interpretation of results due to the lack of agreement on appropriate valuation methodologies and a lack of evidence to support the underlying values of some variables used in the analysis [31].

CBA has been used to evaluate MAR projects with an environmental restoration goal, such as the proposed 6.06 Mm^3/d (1.6 billon US gal/d) ASR system for the Comprehensive Everglades Restoration Plan (CERP) in Florida (USA) [32,33]. Although explicit legal requirements to return damaged ecosystems to baseline functioning may be desirable from an ecological perspective, from an economic perspective it is important to know whether restoration costs generate environmental benefits of equal or greater magnitude [34]. The challenge lies in providing a defensible monetary evaluation of ecosystem services and, for water projects, how those services are affected by variations in water supply. There is a school of thought that CBA, particularly as it is widely applied, is not appropriate because it fails to adequate consider environmental costs and values (*i.e.*, externalities). It has been proposed that ecosystems, such as wetlands, have an existence value, which can be derived simply from the satisfaction of knowing that some feature of the environment continues to exist [24,35]. Ecosystems are also considered to have an intrinsic value, irrespective of the utility or the desires of humans, which lies beyond the scope of CBA.

COSTS OF MAR PROJECTS

The costs of MAR projects include both capital, operations and maintenance costs, and finance costs (debt service). Capital costs are fixed, one-time expenses incurred during the design and construction of the MAR system. Capital costs include, but are not limited to:

- Land;
- Testing costs, feasibility analyses;
- Consulting services for the design, permitting, and supervision of the construction;
- Construction costs (e.g., roads, piping, instrumentation, controls, and pretreatment systems); and
- Regulatory testing requirements during construction and operational testing.

Operation and maintenance costs include the following:

- Labor (system operation, regulatory requirements, administration);
- Electricity;
- Consulting services;
- Regulatory testing requirements (e.g., water quality testing);
- Maintenance costs (e.g., parts replacement, well and basin rehabilitation);
- Pre-treatment costs (additional treatment prior to recharge);
- Post-treatment costs (e.g., chlorination); and
- Raw water costs.

Costs used in the CBA should be marginal not average costs. Sunk costs, which are costs that would be incurred whether a project proceeds or not, should not be included in the CBA. Sunk costs include items such as previously performed hydrogeological investigations, existing wells that are no longer used, and existing intakes and piping. The marginal operational labor cost is zero if existing plant staff can operate the system (*i.e.*, there is no increase in total labor costs). Labor costs are included in the CBA if additional staff (or contracted labor) are needed to operate and manage the system.

CBAs should consider opportunity costs associated with land. Opportunity costs are the benefits one could have received by taking an alternative action. In the case of land, it could be revenues that could have been obtained if the property was sold or rented, or the value of goods and services that would have been obtained if the land were put to an alternative use. MAR systems that utilize wells have minimal surface footprints and, if carefully sited, do not preclude other land uses. Therefore, the opportunity costs associated with MAR systems using wells may be negligible.

The cost of water stored in a potable water ASR system is the marginal cost to abstract and treat the additional recharged water by a water treatment plant, rather than the average production cost or the price charged to customers. Average water costs includes labor, depreciated capital costs, and finances costs (*i.e.*, sunk costs), which would be incurred whether or not the additional water was treated. Local water utilities may obtain water from wholesaler on a take-or-pay basis, in which case they pay for water not used during low demand periods. Hence, there may be a strong financial incentive to store water during low demand periods as the utility is paying for it anyways [34]. In the case of a take-or-pay contract situation, the cost of water would be considered a sunk cost if the water would still be paid for if not used.

The storage space in an aquifer is another potential cost, which is rarely priced in accordance with its scarcity value [36]. Inasmuch as MAR is in its initial stage of development in many areas, there is a low demand for storage space, and it thus has minimal monetary value. However, if MAR implementation locally increases and a scarcity of aquifer storage space with suitable hydrogeologic conditions develops, then one can envision the cost of storage space becoming a significant component of CBA.

BENEFITS OF MAR SYSTEMS

Water has an economic value only when its supply is scarce relative to its demand. Scarce water takes on value because many users compete for it [20]. The benefits of MAR systems are either additional water being available in times of scarcity, improvement in water quality, or a combination of both. Recharge of water can create a new freshwater resource, such as occurs in some ASR systems in which freshwater is emplaced in a brackish aquifer. MAR can also provide benefits by adding water to storage in an aquifer and thus stabilizing or increasing water levels. The total economic value of the recharged water includes its abstraction value plus *in-situ* (non-use) values derived from groundwater being in place.

Economic value is measured on the basis of substitutability, which can be expressed in terms of willingness to pay (WTP) and willingness to accept compensation (WTA) [29]. WTP is the amount someone would be willing to pay rather than do without a good or service. WTA is the minimum amount of money someone would require to voluntarily forgo a good or service. WTP and WTA may not be the same for a given good. Individuals tend to demand considerably greater monetary compensation to give up things that they already possess than they are willing to pay to acquire the same exact items. WTP is also constrained by a person's income in that wealthy people can afford and may thus be willing to pay more for a good or services than would poor people. The economic value to society of a good or service is the aggregate of the WTP of all individuals.

The economic value of water is not a fixed, inherent attribute of a good or service, but rather depends upon time, circumstances, and individual preferences [8]. The scarcity value of water changes with time, with its value increasing during times of decreased supply or increased demand. An important benefit of groundwater, whether placed through natural or enhanced recharge, is as a buffer against variation in surface water or other supplies [8]. Indeed, several studies have demonstrated that the greatest economic benefit of groundwater lies in the stabilizing of water supplies and avoidance of the economic impacts of shortages [37,38,39,40,41]. Surface freshwater flows should be the first source of water used, because they may otherwise be lost if not used when available. Fresh groundwater should optimally be reserved for strategic use in coping with water scarcity. MAR can enhance the ability of groundwater to play a stabilisation role by increasing the available supply of groundwater. Where global climate changes result in locally drier conditions, or a more viable water supply, then water stored in MAR systems would have an even greater value in the future, which needs to be considered in economic analyses.

A fundamental challenge with quantifying the economic benefits of water projects is that there is seldom a free market with respect to water and observed prices do not exist or fail to reflect its social value [42,43]. Often in both developed and developing countries, subsidization is common where water users do not pay the full cost of the construction and operation of the systems through water rates. Construction costs may have been paid for, at least in part, through general government revenues.

Water utilities are essentially monopolies and consequently price regulation is usually applied to protect the public. Publically owned utilities are usually either under direct governmental control or have an elected board. Privately owned utilities are commonly regulated by a governmental agency that has the authority to approve or deny rate increases. For publically owned water utilities, rates are typically determined to generate sufficient revenues to cover operation and maintenance (O&M) expenses, debt service payments, and capital expenditures financed by rates (as opposed to debt and governmental contributions and subsides). Pricing for privately owned utilities is commonly based on a "cost of service" approach, whereby rates are set to generate sufficient revenues to cover O&M expenses, depreciation, taxes (and tax equivalents) and an approved return on base rate.

From an economic perspective, consumers of water should actually pay the marginal cost of water (*i.e.*, the cost to obtain additional supplies) rather than the average cost [44], which is seldom the case. As is often the case for alternative water supply projects of water utilities, the marginal revenues from the additional supplies are less than the marginal costs, and the system is paid for by revenues from the sale of all water, both new alternative supplies and existing conventional supplies.

In developed countries, the price of water represents a small fraction of the household budgets and is usually given little thought. Water is provided at a much lower cost than what the consumer is willing to pay. The price that consumers pay for water can never exceed and seldom approaches the price that they would be willing to pay rather than go without, so the economic benefits derived from the use of water typically exceed the purchase price [20]. In economic terms, utility customers enjoy a substantial consumer surplus in that the value of the water they receive (in turns of WTP) exceeds the price that they pay for the water.

As a result of the consumer surplus, municipal water demand functions show a low elasticity. Rising prices over time may not significantly lower demands, particularly if real incomes are also rising [42]. Some uses are of great necessity to consumers (e.g., potable use, cooking) and there are no practical substitutes. At its limit, as supply approaches zero, the marginal value of water approaches infinity [42]. For example, strategic storage ASR systems are in various stages of development in some Middle Eastern countries that are highly dependent on desalination for the water supply [11]. In the event of a catastrophic disruption of the desalination facilities, due to a natural event, accident or war, millions of people could be without a water supply. The value of a strategic water supply to meet potable demands in an extreme emergency is inestimable [2], even though the probability of such an event is remote. There is thus a low probability that the strategic storage ASR systems could provide enormous benefits. However, placing a meaningful monetary value on the benefits of avoiding a very low probability catastrophic event is very difficult, because there are no precedents.

Since water is rarely priced at a market-determined scarcity value, comprehensive evaluation of MAR schemes require alternative nonmarket valuation methods [35,43,45]. Shadow pricing is typically used in which values are assigned or observed prices are adjusted to correspond to prices that would prevail in a competitive market. Shadow pricing is required, for example, to incorporate environmental costs and benefits in CBA of water projects. Some of the common methods to calculate or estimate the benefits of the water that might be supplied or treated by MAR projects are summarized below (Table 1).

Table 1. Methods to monetize benefits of managed aquifer recharge (MAR) systems

Method	Description
Market prices	Value of water determined by actual prices set by willing buyers and sellers in a competitive market.
Alternative cost	Value of water storage or treatment is determined from the cost of the least expensive alternative that provides comparable benefits.
Value marginal product	The value of water is quantified from the marginal productivity of water, *i.e.*, the extra value of output that can be obtained from additional applications of water.

Contingent value	Survey-based methods to determine an individual's willingness to pay or willingness to accept compensation for a good or service.
Hedonic property value	Value of water is inferred from market transactions (e.g., real estate sales) that are linked to the value of water.
Defensive behavior	Value of a safe and reliable water supply can be estimated from expenditures to avoid exposure to unsafe water.
Damage cost	Value of water is estimated from damage costs avoided, such as health impacts or drought damage.
In-situ groundwater value	MAR system value is estimated from costs avoided resulting from groundwater being in place, such as pumping and land subsidence costs.

Market Prices

Quantification of the value of water is most straightforward where water is sold in a free market. Much has been written over the past two decades on the merits of free water markets as a means of promoting efficient use of water through pricing mechanisms. The principal objections to an entirely free water market system stems largely from the recognition that water is also a social good and that water trading can have significant third party effects (*i.e.*, externalities).

Market pricing systems result in water being allocated to where it results in the greatest net economic returns. The value of water can be determined from direct observations of transactions between willing buyers and sellers [45]. The spot market price under conditions at a given time is a direct measure of WTP. The limitation of using market pricing to determine the value of water is that there are few unfettered markets and that in the absence of a long-term time series of observations, the method may be of limited value for long-term planning purposes [45].

There are very few instances where free market trading prices have been use to quantify the benefits of MAR projects. One example is an evaluation of MAR in the Murrumbidgee region of New South Wales, Australia, in which the value of water was determined using temporary water trading prices (AU$450/ML) during a drought [46]. The spot market price for water will vary depending upon climatic conditions. Stochastic modeling of rainfall, water scarcity, and thus value of water, might be used to estimate potential future revenues from the sale of water over the operational life of a MAR system.

Alternative Cost Method

The alternative cost method is based on the notion that the maximum WTP for a good or service is not greater than the cost of providing that good or service through some other process or technology. The gross benefit of a project is considered to be the cost of the next higher cost alternative. The costs and benefits of MAR and other water projects would be considered relative to other water management options that would achieve the same goals [4, 5, 6, 45, 47]. The alternative cost method is similar to cost-effectiveness analysis in that it does not involve quantification of benefits for each project, which are considered to be constant for all options.

MAR systems with a storage goal can be compared against other options in terms of the unit cost of water recovered or delivered. Where the goal of the system is long-term storage, MAR systems could also be evaluated against other options in terms of the cost per unit storage capacity, with consideration given to recoverability. The alternative cost method is also appropriate for evaluation of MAR projects with a primary water treatment goal. The cost of systems that take advantage of natural vadose zone and aquifer treatment processes can be compared to the costs of alternative engineered solutions that provide the same water quality improvements

A basic problem with the use of the alternative cost method is that a more expensive alternative can always be conceived, which would produce an inflated estimated project benefit [45]. The analysis should demonstrate that the alternative project might actually be built. It is misleading to compare the cost of a MAR project against that of a much more expensive alternative that would never be built. For example, a MAR project to be used for irrigation water supply would be substantially less expensive than a seawater desalination plant built for that same purpose. However, this would be a misleading comparison as the latter would not likely be built, because of the great expense of the desalinated water relative to the value of irrigated crops.

The alternative cost method is commonly used to evaluate water supply projects in which additional water storage capacity or peak demand period water supply are required. For example, potable water demand in South Florida is greatest during the winter and spring dry season, which also coincides with the peak in tourism and seasonal resident population. Permitting of additional fresh groundwater withdrawals is generally no longer possible. The widespread implementation of ASR in Florida starting in the late 1990's was driven by its being the least expensive option to meet seasonal peak water demands [48]. ASR is a less expensive option than the next less expensive option, which is the construction of brackish groundwater desalination capacity that would not be needed (and would thus be idle) for a large part of the year.

Value Marginal Product and Residual Methods

The value marginal product (VMP) method considers the marginal change in the total value of product with a change in input. The value of water is quantified from the marginal productivity of water, which is the extra value of output that can be obtained from additional applications of water [34,49]. With respect to irrigation, the value of water is the change in income with and without an irrigation project, which is a function of increase in crop yield and crop prices. The marginal productivity of water can be calculated from crop-water production functions, which are empirical functions of crop yield *versus* irrigated water applications [43,45]. The function may be either experimental or based on surveys of water users [45]. Production functions are often a function of numerous variables including soil type, fertility, temperature, rainfall, irrigation practices, crop type, and plant growth stage [20]. It can, therefore, be difficult to distill out the specific contribution of irrigation. The increase in yields attributable to irrigation can be alternatively estimated as the difference between irrigated and dryland farming, assuming all other factors being equal. The VMP for irrigation should not be mistaken for water productivity, which is usually defined as the total value of crops divided by the amount of water applied.

The residual method estimates the value of water as the remainder of net income after all other relevant costs are accounted for. The cost of all non-water inputs are deducted from the estimated total value of production. The residual method is most accurate where water constitutes a significant fraction of the value of the output [45,47]. The residual method can result in large potential errors where water is a relatively minor portion of the total value of the product [45]. The residual method tends to give higher estimated values than other methods and over estimates the value of water if other variables are not included in the analysis [45], which is referred to as the "omitted variable problem." Before and after comparisons (irrigated *versus* non-irrigated land) may ignore other variables that influence incomes [22]. There are also disagreements about whether or not and how to consider owned resources (versus contractual resources) such as land, capital, entrepreneurship, and management [45]. Land values can be obtained from rental and sales market.

Limited data are available on the marginal value of water in agriculture in general, and the reported values show a very wide range. Colby [43] reported estimated values of water in agriculture of USD4 to USD236 per acre foot (USD0.003/m^3 to 0.19/m^3) in the western United States. Hussain *et al.* [50] compiled more recent estimates of the value of agricultural water and documented that average values vary greatly across countries and regions, from as low as USD0.001/m^3 to 0.74/m^3.

The VMP has also been used to estimate the value of water in industrial uses. However, water costs are usually very often only a small fraction of total costs [42]. Water supply cost is thus a secondary decision. As water supply and wastewater disposal costs increase, recycling of water increases. However, scattered studies indicate that industrial water demand is quite inelastic [42].

VMP could be used to evaluate the environmental benefits of MAR, such as the restoration and protection of groundwater-dependent ecosystems, in a manner analogous to valuing water for irrigation use. The value of water would be related to the marginal increase in ecosystem services provided by the additional water. The difficult and contentious issue is monetizing ecosystem services. For example, the impacts of an aquifer recharge scheme on spring flows and wetland hydration can be determined through monitoring and modeling. Assigning a monetary value to the benefits of the increased spring flows and wetland hydration has a much greater uncertainty.

Contingent Value Methods

Contingent value methods (CVM), which are also referred to as expressed preference approaches, are survey-based methods used to determine individuals' WTP or WTA for a good or service. The methods involve asking people directly what they would be willing to pay contingent on some hypothetical change in the future state of the world. With respect to environmental issues, a description of conditions simulating a hypothetical market is presented, to which respondents are asked to express their WTP or WTA for existing or potential conditions not registered in any market [45]. A hypothetical application of the CVM to a MAR project is

> *"Your local water utility has completed an investigation of different options to address the current annual water shortages during the summer dry season. The shortages result in restrictions that curtail outdoors water uses, such as lawn and garden watering. The results of the investigation indicate that a managed aquifer recharge system could be constructed that would provide an additional 1 million m³ of water in the summer. The additional water would reduce the need for water use restrictions to less than once in every ten years. Would you be willing to pay an extra $2 per month on your water bill to pay for the MAR system?"*

The cost of the system could be expressed as a discrete choice or evaluated using an iterative bidding process. In the former case, which is referred to as the dichotomous choice or referendum method, a respondent is asked only whether or not they would be willing to pay a specified amount in a specified manner as a "take it or leave it" decision. Iterative bidding processes involve starting with an initial price and then adjusting it upwards or downwards to determine the maximum WTP. For example, if a respondent indicated that they would be willing to pay an additional $2 per month for the MAR system, then they might next be asked if they would be willing to pay $3 per month, and so on, until they indicated no. Conversely, if the response to the initial price is no, then the price would be incrementally reduced until the respondent indicated yes. It has been documented that the initial bid price used can impact survey results.

CVM are subject to a number of potential biases, which has been discussed at great length in the economics literature and was reviewed with respect to CBA and the value of water and the environment by Boardman *et al.* [22], the National Research Council [8] and Young [45]. A basic limitation of CVM is that people's statements about their preferences may not reveal their true preferences and actual behavior, because statement of a WTP does not involve an actual payment obligation. Due to the hypothetical nature of the process, declared intentions may not be accurate guides as to actual future behavior. The biases could either be unintentional or a strategic behavior. As an example of the latter, someone in favor of a water project may intentionally give an excessively high WTP in order to try to influence the survey results. Similarly, respondents may give a low WTP for a project with the hope that in by doing so they may keep future water rates lower. Strategic behavior may be detected as outliers. Sample bias and non-response biases occur when the respondents do not represent all the stakeholders for a project. Interviewer and neutrality biases occur when the respondent perceives that a particular response is preferred by the interviewer or when the question is framed in a manner that is not neutral.

Hedonic Property Value Method

The hedonic property value method is a revealed preference method in which the valuation of non-market goods and attributes is determined by observing market behavior. Expenditures for market goods are linked to the value of nonmarket goods or attributes. The method assumes that an increment in price due to an increase in one characteristic will equal a buyer's WTP for the characteristic as well as the seller's marginal cost of producing that characteristic [45]. The hedonic property value method requires market data and assumes that market participants are able to recognize differences in characteristics. It is commonly based on real estate transactions. A commonly given example is that the value of a living next to a lake can be determined by comparing the sales price of homes with and without a lakefront. With respect to water, the value of groundwater for irrigation use can be estimated from the difference in price of a unit of land with and without a groundwater right or supply. The hedonic property value method assumes that all other variables are equal. However, with respect to water rights in the western United States, the value of water rights depends upon their security (seniority), water quality, and location of use [43].

Defensive Behavior and Damage Cost Methods

The defensive behavior method is based on the WTP to avoid adverse environmental effects [45]. For example, the value of safe drinking water can be estimated from the amount of money that people would pay to avoid exposure to contaminants, such as by purchasing bottled water. The premise of the method is that a rational person will adopt defensive behavior as long as the value of the damage avoided is greater than the cost of the defensive step.

Benefits of MAR systems can be evaluated in terms of damage costs avoided. For example, the benefits of bank filtration systems to provide safer drinking water in rural areas of developing countries can be evaluated in terms of the costs of disease avoided. The costs of disease includes health care expenditures, lost wages and labor (e.g., farmers not be able to tend their fields), and human suffering and premature death. A challenge in evaluating the benefits of water supply and sanitation systems is monetizing the value of a human life and the effects of sickness [31]. An approach taken to evaluate the latter is to use the product of days of work lost and local wages. Similarly, a benefit of water supply projects may be a reduction in the labor required to obtain water, which is a large burden on women and school-age children in areas of some developing countries.

The benefits of an MAR system for irrigation water supply can similarly be estimated from the costs of crop damage during droughts that would be avoided as a result of the stored water. Such an evaluation would require a statistical (probabilistic) analysis of drought frequency and intensity, associated crop damage, and the economic value of lost crops.

In-Situ Values of Groundwater and MAR

In-situ values include a variety of benefits associated with additional groundwater being in place in an aquifer (*i.e.*, higher groundwater levels), as opposed to benefits associated with the abstraction and use of groundwater. *In-situ* benefits are the objectives of systems that involve aquifer recharge without recovery. Reduction in pumping costs is an often cited example of an *in-situ* value that would be a benefit of MAR. Higher groundwater levels result in less energy required to pump water and thus cost savings. The economic value of an MAR system with respect to pumping costs is a function of the change in water level, decrease in energy required to pump the water, and the energy cost. Pumping cost benefits of an MAR project in a given year (C_t) are estimated as [51]:

$$C_t = P_t L_t W_f \tag{2}$$

where P_t is the pumping cost per volume of water per unit of lift per year; L_t is the cumulative average lift change per unit area (ft); and W_t is the amount of water pumped within the affected area without recharge.

Reichard and Bredehoeft [52, 53] performed an economic analysis of the Santa Clara Valley, California, aquifer recharge system. The system uses infiltration basins to recharge a heavily used alluvial aquifer system. A calibrated groundwater flow model was developed and used to calculate the hydraulic effects of the on-going aquifer recharge system. The energy savings was calculated from the modeled increase in heads, annual abstraction volumes, the energy requirements to lift an 1 acre-foot of water (1232 m^3) 1 foot (0.3 M) using a 100% efficient pump, and an average pump efficiency. The energy requirement for a 100% efficient pump is 1.02 KwH to lift 1 acre acre-ft one foot, which is equivalent to 2.71 KwH to lift 1000 m^3 of water 1 m. The benefits of reduced subsidence per foot of drawdown avoided were calculated using an estimate of the economic impacts of historic subsidence divided by the historic drawdown [52, 53].

RISK AND UNCERTAINTY IN CBAS

Perhaps the most neglected aspect of the economics of MAR is addressing risk and uncertainity in CBAs. Risk and uncertainty are often considered synonymous. However, the term "risk" implies that there is some idea of the probability of various events [24,27]. Uncertainty implies that the probability of future events is not known. Although there are without doubt risks and uncertainty associated with the implementation of MAR, as evidenced by some failed or underperforming systems, the existence of risk and uncertainty in projects is seldom acknowledged [11], much less explicitly incorporated into CBAs.

The principle risk and uncertainty associated with MAR systems is that they may fail to meet performance objectives. System performance depends local upon hydrogeologic conditions, which may turn out to be unfavorable for achieving system goals. Adverse results include:

- Recharge may not result in anticipated changes in aquifer water levels;
- Anticipated additional water may not be available when needed (*i.e.*, system has a poor recovery efficiency;
- Unexpected water quality changes due to fluid-rock interactions (e.g., leaching of arsenic into stored water);
- Well performance problems (e.g., low well capacities, well or formation clogging);
- Excessive infiltration basin clogging;
- Water treatment goals are not achieved; and
- Anticipated demand for water (and associated revenues) may not be realized.

For example, the USD150 million dollar Las Posas Basin ASR system in California is considered a failure as it did not achieve water storage goals [54]. The recharge of an enormous volume of water over operational life of the systems did not result in a corresponding increase in aquifer water levels, and thus the water that was "banked" on paper could never be recovered [11].

Some adverse results may be remedied at an additional cost and thus the systems may still be viable. Arsenic leaching and excessive well clogging may be avoided, for example, by pre-treating the recharged water. The additional costs would result in projects having lesser NPVs, but still being economically viable if the benefits are great enough. Some failed ASR systems provided eventual (salvage) value when the wells were put to alternative uses. For example, the Bonita Springs Utilities (Southwest Florida) potable water ASR system encountered hydrogeological conditions that were unsuitable for ASR. A very high degree of aquifer heterogeneity resulted in excessive migration and mixing of injected water and native groundwater [11]. The ASR well was subsequently put to use as the most productive brackish water supply well for their desalination system.

The main source of risk associated with MAR systems stem from a natural groundwater system being used whose hydraulic and geochemical properties can never be fully characterized. The possiblity thus exists that unexpected adverse conditions may be encountered. The risks associated with MAR systems can be reduced, but never entirely eliminated, through high-quality and more-detailed aquifer characterization [11]. Post-audits of both successful and unsuccessful systems can provide valuable lessons that can be a guide for future implementation of MAR [11].

It would clearly be negligent to assume in any CBA that a 100% favorable result will be obtained, when there is a real potential for poor results. Risk and uncertainty can be incorporated into CBA through an expected value analysis, as reviewed by Boardman *et al.* [22]. The future is characterized in terms of a number of distinct contingencies. To evaluate risks, one has to be able to assign probabilities to the occurrence of each possible contingency. Modeling of risk and uncertainty begins with a set of contingencies that are mutually exclusive and capture the full range of likely variations in the costs and benefits of a project or policy.

For example, the net economic benefits of water storage systems vary with the amount of rainfall (and thus demand for water) and the performance of the system (*i.e.*, how much additional water could be recovered when needed). Rainfall also effects natural recharge (thus aquifer water levels and storage space) and the amount of water available for recharge. The average net benefits can be calculated based on the probabilities of different rainfalls and probabilities of different recovery volumes over the operational life of the system. The basic procedure is to identify all potential contingencies and to assign a probability to each. The sum of probabilities for all of the contingencies is equal to one. Probability of each contingency can be based on historic experience (e.g., rainfall data) or subjective opinions of experts. The expected net benefits (ENB) are calculated as:

$$ENB = \sum P_i(B_i - C_i) \tag{3}$$

where P_i = probability of contingency "i"; and B_i and C_i are the present value of the benefits and costs of contingency "i". Not considering risk and uncertainty biases CBAs by increasing expected benefits.

Evaluation of expected net benefits is generally reasonable when there is a pooling of risk, which will make the actual realized values and costs close to the expected values [22]. A limitation of the net expected value method is that it does not capture relevant concerns about extreme negative outcomes [24], particularly where risk is unpooled. Individuals and organizations are often averse to bad outcomes. A low probability risk of a completely failed system may be unacceptable to a small utility with limited resources. The ENB can be weighted to give higher weight to negative outcomes, in the case of a risk-averse decision maker.

Risk analysis can be performed using Monte Carlos analysis, which involves the following main steps [22]:

- Specification of probability distributions for all important uncertain quantitative assumptions;
- Execution of a trial by taking random values drawn from the distribution for each parameter to arrive at a set of specific values for computing realized net benefits; and
- Repetition of the trial numerous times to produce a large number of realizations of net benefits.

The results of all the realizations are used to determine the probability distribution of net benefits.

Risk and uncertainty will also change as a project proceeds. Large MAR projects are implemented in a phased manner. Data collected during an exploratory well program and pilot testing reduce risk and uncertainty and should be used to re-evaluate project feasibility [11]. An updated CBA can thus be performed before the decision is made to construct a full-scale system.

FINANCE OF MAR PROJECTS

MAR projects are primarily funded in four main manners [2]:

- Revenues from the sale of water;
- Direct assessment (pump tax or assessment based on volume of groundwater used);
- Ad valorem tax on real property; and
- General tax revenues.

MAR projects in developing countries may also be funded by external sources, such as international agencies and non-governmental organizations (NGOs). Water users, the primary beneficiaries of projects, should ideally have responsibility for financing projects, through either water rates, direct assessments or ad valorem taxes. Small-scale projects may be constructed through self labor or some sort of cooperative structure. However, the beneficiaries of economically feasible projects may not have financial resources for projects. Construction costs are up front, while benefits occur in the future. The financial constraints are particularly acute in poor areas of developing countries.

The finance of large water supply and storage projects is often controversial because of a dichotomy between the primary beneficiaries of a project and parties that pay for a project. The dichotomy may work in both directions. Projects are often subsidized in that the direct beneficiaries do not fully pay for a project. Governmental and non-governmental agencies may subsidize MAR projects to vary degrees through:

- Projects financed through general revenues or governmental borrowing;
- Grants or low or no interest loans for utilities; and
- Projects entirely funded and constructed by a governmental agency.

Government projects are often favored where concentrated benefits are received by an influential target group and costs are shared in a diffuse manner by society as a whole [22]. Subsidies are commonly justified in terms of secondary benefits. Agricultural projects, for example, support agricultural communities, not just farmers. Subsidies could be justified to achieve societal goals, such as equity (*i.e.*, access of water to all) and food security. Subsidies are justified when the price of a good does not fully reflect its value [45], but can have the adverse impact of encouraging use in quantities greater than the economically efficient quantity.

On the other hand, the operation of a MAR system may provide broader societal and environmental benefits, but the costs may be borne entirely by the system owner (e.g., local water utility and customers). A project may be economically efficient in terms of its total benefits and costs, but not feasible to the owner, because the owner will not receive sufficient personal benefits to cover costs. In other words, there is not an adequate "business case" to justify the investment in MAR. In this case, some sort of governmental subsidy or other means of financial support from more (ideally all) beneficiaries may be justified.

The issue of finance is well illustrated by the Las Vegas, Nevada (USA) aquifer recharge system, for which Donovan *et al.* [55] provided a cost-benefit analysis for non-municipal water users. The Las Vegas aquifer system recharges an historically overdrafted alluvial aquifer with seasonally available excess treated surface (Colorado River) water, with the primary goals of increasing available water resources, slowing or reducing the decline in water levels, and reducing the rate of land subsidence. The recharge is performed using injection wells and water is abstracted by municipal users and non-municipal water users using privately owned on-site wells.

The net benefits of the system to non-municipal users were calculated to be about USD700 per 1230 m^3/year, which is largely from cost savings from deferment or elimination of the need to rehabilitate and replace wells. Well rehabilitation would consist of deepening wells and/or lowering pumps in response to a continuing decline of aquifer water levels that would otherwise occur without the recharge. There are additional minor savings from reduced energy consumption. Non-municipal users were receiving these benefits for free. The solution was to implement a groundwater management program (GMP) in which non-municipal users are charged on either a per well or permitted water use rate basis in order to support the system.

DISCUSSION

Economic analysis of MAR systems are inherently project specific, depending upon the type of system, performance objectives, local hydrological and physical conditions, planned uses of the recovered and stored water, and alternative water supply and treatment options. General and system-specific feasibility is dictated by their benefits, which is determined by the value of water. ASR and other forms of MAR are usually economically feasible (*i.e.*, have a positive NPV), where water is used for municipal (potable) use in water scarce regions, provided that local hydrogeologic conditions are favorable for achieving system performance goals.

Multiple approaches may be appropriate to evaluate the economics of MAR projects. Consider, for example, a riverbank filtration (RBF) system to improve potable water quality in a developing country. The economic viability of the project could be considered using cost-effectiveness, by comparison of the costs of the RBF system with other options that would provide comparable water quality benefits.

Alternatively, a CBA could be performed in which the present value of the costs of the systems is compared to the present value of the benefits provided. Expected benefits might be a reduction in sickness and premature mortality that are the result of ingestion of or contact with water-borne pathogens. Proposed systems should be economically viable using both approaches.

The actual costs of MAR systems in terms of total costs and cost per volume of recovered or treated water are highly system-specific. In general, MAR systems provide the greatest benefits where the water is put to a high value use and alternative, inexpensive options are not available. Potable water ASR in South Florida provides a good example of some of the economic issues associated with MAR for a high value use. ASR has been implemented primarily to meet peaks in demand during the winter and spring dry season. The current costs for brackish groundwater desalination (the least expensive alternative) are commonly now in the USD0.30/m^3 to 0.60/m^3 range, which is based on full-time operation of the plants. The costs of desalinated water from facilities constructed only to meet peaks in demand would be substantially greater than the above estimates (approximately USD0.70/m^3 to 1.50/m^3), because the large annual depreciated construction and fixed operational costs for a desalination system would be divided by a relatively small seasonal production volume, resulting in higher unit costs. Desalination costs would also depend upon whether an existing plant is expanded or a new plant is constructed, and the size of the plant and associated economy of scale.

There is a substantial economy of scale associated with wells. For example, doubling of the capacity of wells typically involves significantly less than a doubling in the cost of the well, wellhead, pump, and piping. The costs of regulatory compliance are also independent of well capacity. On the benefit side, cost per unit volume is directly proportional to the volume of water recovered, which is a function of system capacity, recovery efficiency, and demand (*i.e.*, amount of available water that is actually recovered). The annual cost for a 1 to 2 million gallons per day (3788 to 7576 m3/d) ASR system that is recovered for 90 days per year is on the order of USD0.30/m^3 to 0.60/m^3 in South Florida. The cost of ASR to meet peaks in potable water demand, therefore, can be 50% or less than the cost of brackish water desalination.

The economics of MAR for irrigation water supply are much more variable because of the wide range of monetary values of water associated with this use. The value of water for irrigation depends upon the crop type being grown, and is typically relatively low for cereal crops and greater for fruits and vegetables. The value of water in agriculture also depends upon local market prices for crops. A wide range of values has been presented for the value of water in agriculture with most being no more than USD0.001/m^3 to 0.79/m^3 [50]. Hence, MAR systems for agricultural water supply need to be low cost and passive (*i.e.*, do not require large amounts of energy and human intervention to operate). MAR methods most appropriate for irrigation water supply are systems that recharge untreated water (stormwater and flood water) using infiltration basins and ponds, and in-channel modifications (e.g., check dams).

Small-scale MAR systems for potable water supply are right-sized for some rural areas and developing countries. For example, production of water from riverbank filtration systems consisting of drilled or dug shallow wells located adjacent to a river can be a very cost-effective means to improve water quality with concomitant health benefits. Riverbank filtration has been demonstrated to be a less expensive option than conventional surface water intakes and filtration systems where local hydrogeologic conditions are favorable. One small-scale MAR application in India is the inexpensive retrofitting of existing tube wells to allow for aquifer recharge whenever excess rain or canal water is available [56]. The main cost elements are construction of a connecting channel to convey canal water and construction of a settling basin and filter tank [57]. A variety of other methods are employed in India to enhance recharge such as surface spreading using percolation tanks (ponds) and check dams constructed across or near streams, and drainage channels in order to impound runoff and retain it for a longer time to increase the opportunity time for recharge [58].

Managed aquifer recharge of recycled or reclaimed water can be a valuable water resources management tool as it may allow for more of this resource to be put to beneficial use and avoids the costs and environmental impacts of its disposal. For example, the primary economic benefit of a reclaimed water ASR system is Destin, Florida (USA), is that it is much less expensive there to store excess reclaimed water underground during periods when supplies exceed demands than to construct new disposal facilities due to limited land availability and regulatory and political objections to an offshore outfall [59]. The ASR system also has the important benefit of increasing the reliability of the reclaimed water supply, which makes potential customers more willing to commit to connecting to the reuse system. However, a widespread constraint on the implementation of reclaimed water ASR is that the water is often provided to customers at a low (and in some instances no) cost, so there is little financial incentive (benefits received) to invest in the systems, particularly where a low-cost disposal option is available.

An important area for additional research is the collection and analysis of accurate data on the economics of existing MAR systems. Data are needed on the construction and operational costs and benefits of the various types of systems and in different geographic locations. The conceptual framework exists for evaluation of the economics of MAR systems, but there is a paucity of hard data to perform meaningful cost-benefit analyses. The paucity of actual data on the economics of MAR systems, which demonstrates that their benefits exceed costs, is a continued impediment to the further implementation of the technology.

CONCLUSIONS

The economic feasibility of MAR can be evaluated using conventional CBA in which the NPV of system options are determined and compared against each other and other water storage and treatment options. The CBA process should be rigorous and consider all marginal costs and benefits, risks, and opportunity costs. The greatest uncertainty in CBA analysis of MAR relates to monetizing benefits, which ties into the more basic question of the value of water. In the absence of a free market-derived WTP price for water, shadow pricing is required to estimate project benefits, such alternative cost and value marginal product methods. A major deficiency of past economic analyses of MAR is the failure to consider risk, particularly the effect of possible system under-performance in reducing system benefits.

CBA of MAR systems is highly dependent on site-specific conditions. In general, systems are economically viable where the water is put to a high-value use, such as potable and some industrial and irrigation water supplies. MAR system for lower value irrigation water supply (e.g., cereal crops) should be low cost, passive systems. MAR systems should ideally be financed by the primary project beneficiaries. As is the case for many water projects in general, MAR projects are often subsidized when beneficiaries are unable or unwilling to pay the full costs. Finance of MAR can be particularly challenging in rural areas of developing countries where financial resources are limited and the construction costs have to be borne before benefits of the systems are realized.

ACKNOWLEDGMENTS

The author thanks Scott Manahan, Andrew Ross, and three anonymous reviewers for their thoughtful reviews.

REFERENCES

1. Dillon, P. Future management of aquifer recharge. Hydrogeol. J. 2005, 13, 313–316. CrossRef
2. Todd, D.K. Economics of ground water recharge. J. Hydraul. Div.-ASCE 1965, 91, 249–270.
3. Botzan, T.M.; Necula, A.I.; Mariño, M.A.; Basağaoğlu, H. Benefit-cost model for an artificial recharge scenario in the San Jaoquin Valley, California. Water Resour. Manag. 1999, 13, 189–203.
4. Frankel, R.J. Economics of artificial recharge for municipal water supply. In Proceedings of the Artificial Recharge and Management of Aquifers, Symposium of Haifa, Haifa, Israel, 19–26 March 1967; IAHS Press: Wallingford, UK, 1967; pp. 289–301.
5. Khan, S.; Mushtaq, S.; Hanjra, M.A.; Schaeffer, J. Estimating potential costs and gains from an aquifer storage and recovery program in Australia. Agric. Water Manag. 2008, 95, 477–488.
6. Cowdin, S.W.; Peters, H.J. The Economics of Ground Water Recharge Projects. In Artificial Recharge of Groundwater, Proceedings of the International Symposium on Artificial Recharge of Groundwater, Anaheim, CA, USA, 23–27 August 1988; Johnson, A.I., Finlayson, D.J., Eds.; American Society of Civil Engineers: New York, NY, USA, 1989; pp. 14–22.

7. Bergstrom, J.C.; Boyle, K.J.; Job, C.A.; Kealy, M.J. Assessing the economic benefits for ground water policy decisions. Water Resour. Bull. 1996, 32, 279–291.
8. National Research Council. Valuing Ground Water, Economic Concepts and Approaches; National Academy Press: Washington, DC, USA, 1997.
9. Huisman, L.; Olsthoorn, T.N. Artificial Groundwater Recharge; Pitman Advanced Publishing: Boston, MA, USA, 1983.
10. Pyne, R.D.G. Aquifer Storage Recovery: A Guide to Groundwater Recharge through Wells; ASR Systems: Gainesville, FL, USA, 2005.
11. Maliva, R.G.; Missimer, T.M. Aquifer Storage and Recovery and Managed Aquifer Recharge Using Wells: Planning, Hydrogeology, Design, and Operation; Schlumberger Corporation: Sugar Land, TX, USA, 2010.
12. Oaksford, E.T. Artificial recharge: Methods, hydraulics, and monitoring. In Artificial Recharge of Groundwater; Asano, T., Ed.; Butterworth Publishers: Stoneham, MA, USA, 1985; pp. 69–127.
13. Roscoe Moss Company. Handbook of Ground Water Development; Wiley: New York, NY, USA, 1990.
14. Bouwer, H. Ground water recharge with sewage effluent. In Artificial Recharge of Groundwater, Proceedings of the International Symposium on Artificial Recharge of Groundwater, Anaheim, CA, USA, 23–27 August 1988; Johnson, A.I., Finlayson, D.J., Eds.; American Society of Civil Engineers: New York, NY, USA, 1989; pp. 170–185.
15. Bouwer, H. Groundwater recharge with sewage effluent. Water Sci. Technol. 1991, 23, 2099–2108.
16. Pescod, M.E. Wastewater Treatment and Use in Agriculture; FAO Irrigation and Drainage Paper 47. Food and Agriculture Organization of the United Nations: Rome, Italy, 1992.
17. Fox, P.; Houston, S.; Westerhoff, P. Soil Aquifer Treatment for Sustainable Water Supply; American Water Works Association Research Foundation: Denver, CO, USA, 2001.
18. Fox, P.; Houston, S.; Westerhoff, P.; Drewes, J.E.; Nellor, M.; Yanko, W.; Baird, R.; Rincon, M.; Arnold, R.; Lansey, K.; et al. Investigations of Soil-Aquifer Treatment for Sustainable Water

Reuse; National Center for Sustainable Water Supply: Tempe, AZ, USA, 2001.

19. Rinck-Pfeiffer, S.; Pitman, C.; Dillon, P. Stormwater ASR in practice and ASTR under investigation in Salisbury, South Australia. In Recharge Systems for Protecting and Enhancing Groundwater Resources, Proceedings of the 5th International Symposium on Management of Aquifer Recharge, Berlin, Germany, 11–16 June 2005; UNESCO: Paris, France; pp. 151–159.
20. Ward, F.A.; Michelsen, A. The economic value of water in agriculture: Concepts and policy applications. Water Policy 2002, 4, 423–466.
21. Layard, R.; Glaister, S. Cost-Benefit Analysis, 2nd ed.; Cambridge University Press: Cambridge, UK, 1994.
22. Boardman, A.; Greenberg, D.; Vining, A.; Weimer, D. Cost-Benefit Analysis: Concepts and Practice; Prentice Hall: Upper Saddle River, NJ, USA, 1996.
23. Hanley, N.; Spash, C.L. Cost-Benefit Analysis and the Environment; Edward Elgar: Cheltenham, UK, 1993.
24. Turner, R.K.; Pearce, D.; Bateman, I. Environmental Economics: An Elementary Introduction; Johns Hopkins Press: Baltimore, MD, USA, 1993.
25. Pearce, D.; Atkinson, G.; Mourato, S. Cost-Benefit Analysis and the Environment: Recent Developments; Organization for Economic Co-operation and Development: Paris, France, 2006.
26. Atkinson, G.; Mourato, S. Environmental Cost-Benefit Analysis. Annu. Rev. Environ. Resour. 2008, 33, 317–344.
27. Turner, K.; Geprgiou, S.; Clark, R.; Brouwer, R.; Burke, J. Economic Evaluation of Water Resources in Agriculture: From the Sectoral to a Functional Perspective of Natural Resource Management; FAO Water Report 27. Food and Agricultural Organization of the United Nations: Rome, Italy, 2004.
28. Reisner, M. Cadillac Desert: The American West and Its Disappearing Water, 2nd ed.; Penguin: New York, NY, USA, 1999.
29. Freeman, A.M., III. The Measurement of Environmental and Resource Values; Resources for the Future: Washington, DC, USA, 1993.

30. Kolstad, C.D. Environmental Economics; Oxford University Press: New York, NY, USA, 2000.

31. Hutton, G.; Haller, L.; Bartram, J. Global cost-benefit analysis of water supply and sanitation interventions. J. Water Health 2007, 5, 481–502.

32. Milon, J.W.; Hodges, A W.; Rimal, A.; Kiker, C.F.; Casey, F. Public Preferences and Economic Values for Restoration of the Everglades/South. Florida Ecosystem; University of Florida, Food & Resource Economics Department, Florida Agricultural Experiment Station, Institute of Food and Agricultural Sciences: Gainesville, FL, USA, 1999.

33. Mather Economics. In Measuring the Economic Benefits of America's Everglades Restoration; Everglades Foundation: Palmetto Bay, FL, USA, 2010.

34. Holl, K.D.; Howarth, R.B. Paying for restoration. Restor. Ecol. 2000, 8, 260–267.

35. Turner, R.K.; van den Bergh, J.C.; Söderqvist, T.; Barendregt, A.; van der Straaten, J.; Maltby, E.; van Ierland, E.C. Ecological-economic analysis of wetlands: Scientific integration for management and policy. Ecol. Econ. 2000, 35, 7–23.

36. National Research Council. Prospects for Managed Underground Storage of Recoverable Water; National Academies Press: Washington, DC, USA, 2008.

37. Tsur, Y. The stabilization role of groundwater when surface water supplies are uncertain: The implications for groundwater development. Water Resour. Res. 1990, 26, 811–818.

38. Purkey, D.R.; Thomas, G.A.; Fullerton, D.K.; Moench, M.; Axelrad, L. Feasibility Study of a Maximal Program of Groundwater Banking; Natural Heritage Institute: San Francisco, CA, USA, 1998.

39. Gemma, M.; Tsur, Y. The stabilization value of groundwater and conjunctive water management under uncertainty. Appl. Econ. Perspect. Policy 2007, 29, 540–548.

40. Reichard, E.G.; Li, Z.; Hermans, C. Emergency use of groundwater as a backup supply: Quantifying hydraulic impacts and economic benefits. Water Resour. Res. 2010, 46.

41. Reichard, E.G.; Raucher, R.S. Economics of conjunctive use of groundwater and surface water. In Water: Science, Policy, and Management; Water Resources Monograph 16. Lawford, R., Fort, D., Hartmann, H., Eden, S., Eds.; American Geophysical Union: Washington, DC, USA, 2003; pp. 161–176.
42. Gibbons, D.C. The Economic Value of Water; Resources for the Future: Washington, DC, USA, 1986.
43. Colby, B.G. Estimating the value of water in alternative uses. Nat. Resour. J. 1989, 29, 511–527.
44. Winpenny, J. Managing Water as an Economic Resource; Rutledge: London, UK, 1994.
45. Young, R.A. Determining the Economic Value of Water Concepts and Methods; Resources for the Future: Washington, DC, USA, 2005.
46. Harou, J.J.; Lund, J.R. Ending groundwater overdraft in hydrologic economics systems. Hydrogeol. J. 2008, 16, 1039–1055.
47. Agudelo, J.I. The Economic Valuation of Water: Principles and Methods; Value of Water Resources Report 5. IHE Delft: Delft, The Netherlands, 2001.
48. Maliva, R.G.; Missimer, T.M. Cost-benefit analyses of aquifer storage and recovery. In Florida—The Reality beyond the Hype; Proceedings of the 2007 Florida Water Resources Conference, Orlando, FL, USA, 15–18 April 2007.
49. Young, R.A. Measuring Economic Benefits for Water Investments and Policies; World Bank Technical Paper No. 338. World Bank: Washington, DC, USA, 1996.
50. Hussain, I.; Turral, H.; Molden, D. Measuring and enhancing the value of agricultural water in irrigated river basins. Irrig. Sci. 2007, 25, 263–282.
51. Supalla, R.J.; Comer, D.A. The economic value of ground water recharge for irrigation use. Water Resour. Bull. 1982, 18, 679–686.
52. Reichard, E.G.; Bredehoeft, J.D. An engineering economic analysis of a program for artificial recharge. Water Resour. Bull. 1984, 20, 929–939.
53. Reichard, E.G.; Bredehoeft, J.D. Incorporating ground water modeling into cost-benefit analyses or artificial recharge. In

Proceedings of the NWWA Conference of Ground Water Management, Orlando, FL, USA, 29–31 October 1984; pp. 343–355.

54. Blood, M.R.; Spagat, E. Las Posas Basin Aquifer Failure Illustrates Risks of Underground Reservoirs; Associated Press: New York City, USA, 2013.
55. Donovan, D.J.; Katzer, T.; Brothers, K.; Cole, E.; Johnson, M. Cost-benefit analysis of artificial recharge in Las Vegas Valley, Nevada. J. Water Resour. Plan. Manag. 2002, 128, 356–365.
56. Malik, R.S.; Jhorar, B.S.; Jhorar, R.K.; Streck, T.; Richter, J. Long-term successful operation of existing brackish cavity wells for ASR to improve quality for irrigation by Indian farmers. In Management of Aquifer Recharge for Sustainability; Dillon, P.J., Ed.; A.A. Balkema: Lisse, The Netherlands, 2002; pp. 465–468.
57. Goyal, V.; Jhorar, B.S.; Malik, R.S.; Streck, T. Performance evaluation of aquifer storage recovery wells for conjunctive water management as influenced by buffer storage volume and storage time. Curr. Sci. 2008, 94, 465–472.
58. Sakthivadivel, R. The groundwater recharge movement in India. In The Agricultural Groundwater Revolution: Opportunities and Threats to Development; Giordano, M., Villholth, K.G., Eds.; CAB International: Wallingford, UK, 2007; pp. 195–210.
59. Maliva, R.G.; Griswold, R.F.; Autrey, M.M. Prototype for a reclaimed water aquifer storage recovery system benefits and operational experiences. Fla. Water Resour. J. 2013, 65, 54–59.

Chapter 7

The Role of Transdisciplinary Approach and Community Participation in Village Scale Groundwater Management: Insights from Gujarat and Rajasthan, India

Basant Maheshwari [1], Maria Varua [1], John Ward [2], Roger Packham [1], Pennan Chinnasamy [3], Yogita Dashora [4], Seema Dave [1], Prahlad Soni [5], Peter Dillon [2], Ramesh Purohit [4], Hakimuddin [5], Tushaar Shah [3], Sachin Oza [6], Pradeep Singh [4], Sanmugam Prathapar [3], Ashish Patel [6], Yogesh Jadeja [7], Brijen Thaker [7], Rai Kookana [2], Harsharn Grewal [1], Kamal Yadav [4], Hemant Mittal [4], Michael Chew [1] and Pratap Rao [4]

[1]School of Science & Health, University of Western Sydney, Locked Bag 1797, Penrith NSW 2751, Australia

[2]CSIRO Land and Water Flagship, Glen Osmond, SA 5064, Australia

[3]International Water Management Institute, PO Box 2075, Colombo, Sri Lanka

[4]Department of Soil & Water Conservation Engineering, Maharana Pratap University of Agriculture and Technology, Udaipur, Raj. 313001, India

[5]Vidya Bhawan Krishi Vigyan Kendra, Udaipur, Raj. 313001, India

[6]Development Support Centre, Ahmedabad, Guj. 380058, India

[7]Arid Communities and Technologies, Bhuj, Guj. 370001, India

ABSTRACT

Sustainable use of groundwater is becoming critical in India and requires effective participation from local communities along with technical, social, economic, policy and political inputs. Access to groundwater for farming communities is also an emotional and complex issue as their livelihood and survival depends on it. In this article, we report on transdisciplinary approaches to understanding the issues, challenges and options for improving sustainability of groundwater use in States of Gujarat and Rajasthan, India. In this project, called Managed Aquifer Recharge through Village level Intervention (MARVI), the research is focused on developing a suitable participatory approach and methodology with associated tools that will assist in improving supply and demand management of groundwater. The study was conducted in the Meghraj watershed in Aravalli district, Gujarat, and the Dharta watershed in Udaipur district, Rajasthan, India. The study involved the collection of hydrologic, agronomic and socio-economic data and engagement of local village and school communities through their role in groundwater monitoring, field trials, photovoice activities and education campaigns. The study revealed that availability of relevant and reliable data related to the various aspects of groundwater and developing trust and support between local communities, NGOs and government agencies are the key to moving towards a dialogue to decide on what to do to achieve sustainable use of groundwater. The analysis of long-term water table data indicated considerable fluctuation in groundwater levels from year to year or a net lowering of the water table, but the levels tend to recover during wet years.

This provides hope that by improving management of recharge structures and groundwater pumping, we can assist in stabilizing the local water table. Our interventions through *Bhujal Jankaar*s (BJs), (a Hindi word meaning "groundwater informed" volunteers), schools, photovoice workshops and newsletters have resulted in dialogue within the communities about the seriousness of the groundwater issue and ways to explore options for situation improvement. The BJs are now trained to understand how local recharge and discharge patterns are influenced by local rainfall patterns and pumping patterns and they are now becoming local champions of groundwater and an important link between farmers and project team. This study has further strengthened the belief that traditional research approaches to improve the groundwater situation are unlikely to be suitable for complex groundwater issues in the study areas. The experience from the study indicates that a transdisciplinary approach is likely to be more effective in enabling farmers, other village community members and NGOs to work together with researchers and government agencies to understand the groundwater situation and design interventions that are holistic and have wider ownership. Also, such an approach is expected to deliver longer-term sustainability of groundwater at a regional level.

INTRODUCTION

India is the largest user of groundwater in the world with an estimated usage of 230 km3 per year [1]. Globally, areas under groundwater irrigation are the highest in India (39 million ha), followed by China (19 million ha) and the USA (17 million ha), and at present 204 km3 y−1 of groundwater is pumped annually in India [2]. Several reasons may be attributed to this phenomenon. Access to groundwater increased since the 1970s, when diesel and electric pumps became affordable to most small landholders. The causes of increased groundwater use are also rooted in population growth and economic expansion, and as result the annual groundwater use now probably exceeds the annual rainfall recharge. The notion of groundwater as a private resource, the rights of which are associated with land rights, has led to an exploitative extraction regime [3].

Farmers in semi-arid parts of India use groundwater to save rainfed crops from failure and to increase yields. As it is a relatively cheap and easily accessible water resource for individual farmers, irrespective of their farm size, groundwater is often extracted beyond its natural recharging capacity. With increased use of groundwater, the depth to the water table in many parts are fluctuating considerably during the year and the use of groundwater has risen to a level that groundwater from shallow aquifers is not adequate to meet the rising demand. Hence, groundwater from deeper aquifers is being pumped by the drilling of tube wells. There are also instances where fresh groundwater at shallow depths has been depleted, rendering marginal quality water from deeper layers of the aquifer [4]. The extensive use of groundwater resources by farmers all over the country pumping out water in an unregulated manner creates its own sets of complex management and sustainability issues.

The use of groundwater in agriculture is important in India, as it has enabled farmers to manage deficiencies in monsoonal rainfall, allowed dry-season irrigation, thus contributing to poverty alleviation. For this reason, a range of on-ground works to recharge groundwater are being implemented at the village scale throughout India as a part of the Government of India's "Mahatma Gandhi National Rural Employment Guarantee Act" (MNREGA) to enhance livelihood opportunities while developing a durable asset base. A significant part of the investment through MNREGA is for enhancing long-term, local water security by on-ground structures such as check dams, percolation tanks, surface spreading basins, pits and recharge shafts [5]. The development of on-ground structures to enhance groundwater recharge in India is called "watershed development". It is a long running program of Government of India and has significant hydrologic consequences, in particular, altering the runoff regime in downstream regions and groundwater recharge at local and regional scales.

In spite of all the efforts in the past to improve the sustainability groundwater in India, the problem of groundwater management is still severe, particularly in Rajasthan and Gujarat. In this project, called Managed Aquifer Recharge through Village level Intervention (MARVI), the research is focused on developing a suitable participatory approach and methodology with associated tools that will assist in improving supply and demand management of groundwater. Another important aspect of the project is education of and engagement with village communities, local NGO and government agencies to facilitate them working together to achieve sustainable groundwater management. In this article, we report on some key findings from the MARVI project with two main objectives: (i) to show how basic hydrologic information collected by farmers and supplemented with hydrologic, agronomic and socio-economic data collected by the project team is leading to an assessment and understanding of the groundwater storage changes; and (ii) to reveal how this information and engagement activities can be used to empower village communities and other stakeholders to develop and assess their own viable options for groundwater management, including managed aquifer recharge and measures to reduce water demand while sustaining livelihoods.

THE STUDY WATERSHEDS

The work reported here was conducted in the Meghraj watershed in Aravalli district, Gujarat, and the Dharta watershed in Udaipur district, Rajasthan, India (Figure 1). Both watersheds have a semi-arid climate, with the average annual rainfall in excess of 600 mm, but more than 90% of this rainfall is received during the monsoon months of June to September. Most farmers in the two watersheds grow maize, black gram, mungbean, guar, soybeans (recently introduced) and vegetables as Kharif crops during the rainy season. Wheat, gram and mustard are the main Rabi crops grown during the winter season. Farmers who have access to groundwater (and in some instances canal water) grow two crops a year and those who have access to water supplies throughout the year also grow some summer crops such as vegetables and fodder.

The occurrence and distribution of rainfall in both the Meghraj and Dharta watersheds are highly uneven in both time and space. Kharif crops are mainly dependent on the vagaries of the monsoon and are often at risk of either complete or partial crop failure due to inadequate rainfall, or rainfall not occurring at a critical stage of crop growth. Therefore, the uneven and erratic distribution of rainfall provides a major challenge to growing crops successfully and to sustaining a decent livelihood. When rainfall does not occur at the right time or in the required amount, some supplementary irrigation, also called "life saving irrigation", using rainwater stored on the surface or drawn from the underground aquifer systems can make a huge difference in avoiding crop failure.

Figure 1. The Meghraj and Dharta watersheds. The inset map shows the location of the watersheds in the states of Gujarat and Rajasthan in India.

A number of in situ conservation measures, including farm ponds, percolation ponds and check dams have been constructed in the two watersheds under both the Integrated Watershed Management (IWM) programs and MNREGA. The State Governments of Gujarat and

Rajasthan, along with the Central Government have invested significant amounts in these two watersheds in the past, and continue to do so by constructing more of these structures. However, it is not clear how effective these programs are, and what impacts these investments are having on groundwater security. Figure 2 and Figure 3 show the MAR structures in the Meghraj and Dharta watersheds.

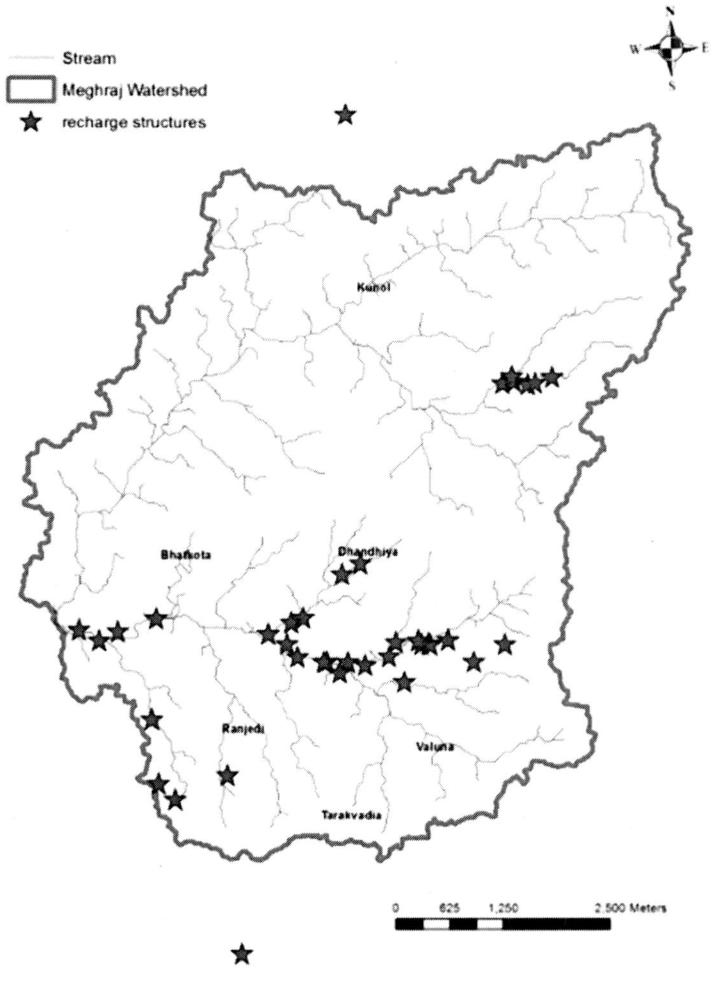

Figure 2. Location of MARs in Meghraj.

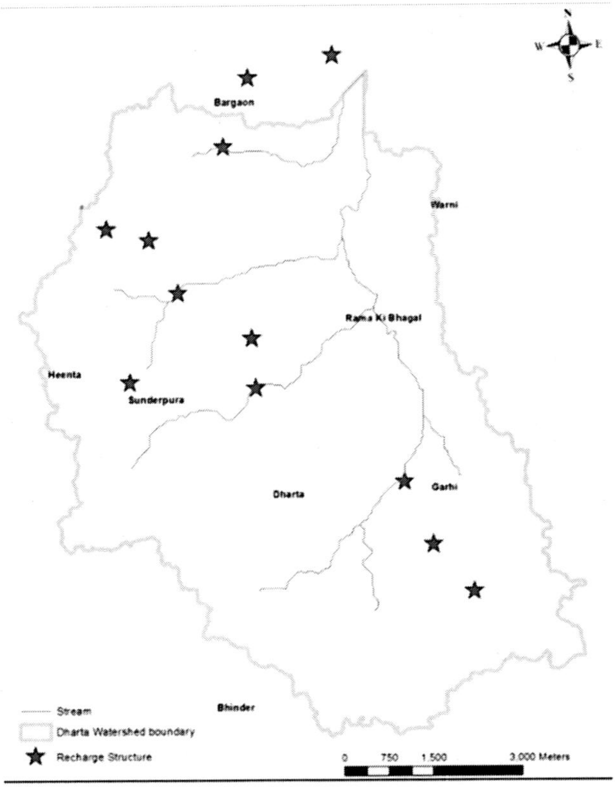

Figure 3. Location of MARs in the Dharta watershed.

It is important to note that both watersheds are in hard rock aquifer areas. It well known that hard rock aquifers have low porosity and low connectivity and the movement of groundwater occurs through faults, fissures and fractures. Hence they store limited volumes, and when stored water is withdrawn by pumps, the emptied pores are not immediately filled by flows from adjacent areas. As result of low rain-recharge, and low porosity and low connectivity, the depth to water table fluctuates considerably during the year and significant water scarcity is often experienced during summer months or drier years.

Most farmers in the Meghraj watershed belong to a tribal community, while those in the Dharta watershed are from mainstream groups. The farming practices in the two watersheds have not advanced adequately to cope with declining water supplies. For this reason, the physical and socio-economic conditions in the two watersheds provide a diversity of transdisciplinary research opportunities and engagement issues around groundwater recharge and management.

STUDY APPROACH

The study approach in the MARVI project is underpinned by transdisciplinary research with a main focus at the "village scale" to understand the complex interrelations between rainfall, aquifer recharge, groundwater pumping and livelihood opportunities. We define transdisciplinary research as one in which both researchers from different unrelated disciplines and non-academic participants, such as farmers and other villagers, work together for a common goal and create new knowledge and theory to improve a complex situation. Thus, in this project we recognized the importance of involving local villagers and other stakeholders through this approach during the research process and engaged them in participatory groundwater monitoring and education to explore options for groundwater sustainability. Figure 4 illustrates the application of relevant social and natural sciences research and engagement to improve the field situation.

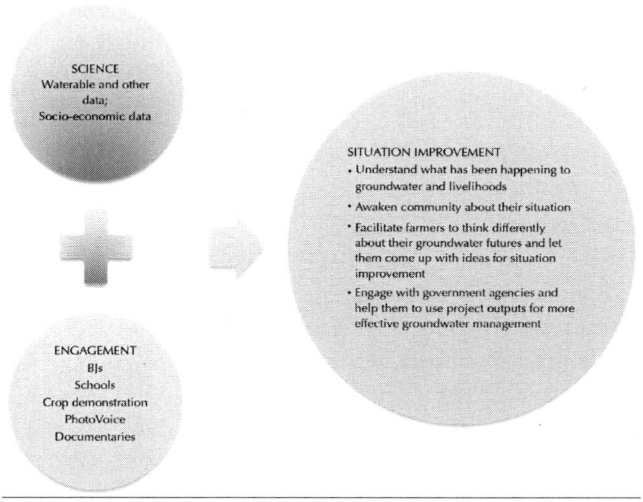

Figure 4. Study approach in MARVI project.

With an active engagement of local villagers, the project team collected a range of hydrologic, agronomic, economic, social and cultural data at selected clusters of villages over a two-year period. The engagement of villagers and data collected are then employed to understand the current situation and develop bio-physical and socio-economic insights to evaluate the current issues, identify options and strategies, provides a scientific and evidence-based input to enhance watershed development policies.

FIELD RESEARCH AND DATA ANALYSIS

Participatory Groundwater Monitoring

A desired outcome of the MARVI project would be collective action at village level that is mutually beneficial to all the villagers, and from which other communities could learn. To achieve this, [6] have shown the need to develop Social Capital. This project used participatory approaches to help to develop social capital competences, with training programs aimed at supporting cognitive aspects of this social capital competence. In addition, the project used participatory monitoring for some data collection to also support this development.

Participatory monitoring of the water table was achieved through the engagement of villagers in the two watersheds. A total of nine local villagers, called *Bhujal Jankaars* (BJs)-a Hindi word meaning "Groundwater Informed" volunteers—were recruited in the Meghraj watershed and similarly 25 BJs were selected in the Dharta watershed. The main idea of recruiting BJs into this project was to give local villagers ownership in the project, build their capacity so that they can understand their groundwater issues and eventually help them to become champions of their community for improving the groundwater situation. The BJs were trained in a number of relevant aspects, such as mapping, water table and water quality measurements. They were also exposed to basic hydro-geologic concepts influencing groundwater availability for agricultural use.

The BJs were involved in weekly monitoring of the water table in open wells, 110 wells in the Meghraj watershed and 250 in the Dharta watershed. Prior to the monitoring of wells, all BJ's did a baseline survey with the help of the project team to compile the required information about village wells. The BJs monitored the groundwater changes through the measurement of water level depth from the ground surface on weekly basis and pH and EC on monthly basis. To assist in the reliability of the data collected by BJs, the project staff each week randomly measured the water level depth data in some of the wells using the same method as those of the BJ and crosschecked water level depths with those measured by BJs. This ensured that BJs were collecting the data properly.

Hydrologic Measurements

Two automatic weather stations, one in each watershed, were installed to collect local weather information for water balance modeling and evaluating the effectiveness of recharge structures on groundwater levels. In addition, six automatic rain gauges were installed in local schools in the Meghraj watershed and five in the Dharta watershed. The purpose of engaging schools in rainfall measurements was to make the school children aware of the water availability in the area and its importance. Some villagers, acting as BJs in the two watersheds, were also given manual rain gauges to monitor rainfall.

A total of five groundwater depth sensors were installed in the Dharta watershed and three in the Meghraj watershed for monitoring water table depth at 15 min intervals. The measurement of water table depth at such a short interval is helpful to analyze rapid changes in water table depth following a pumping event or significant rainfall occurrence. Four water meters in each watershed were installed to measure pumped volume and water productivity for specific crops. Groundwater and soil samples were collected in the watersheds at different times during the study to examine whether they impose limitations for crop production and consequently on the livelihood of people.

The Central Ground Water Board (CGWB), an Indian Government organization, maintains and monitors observation wells across the country. In Gujarat and Rajasthan, CGWB monitors 1197 and 1111 wells, respectively [7]. The data is collected four times, Post-Monsoon (*Rabi*), Pre-Monsoon, Monsoon and Post-Monsoon (*Kharif*), which correspond to January, May, August and November respectively. For the current study, we chose two wells that fall in our study watershed areas. The data was collected from the WRIS website [8] which is maintained by the Indian Space Research Organisation (ISRO) and Central Water Commission (CWC).

Socio-Economic Survey

Households in the two watersheds that contributed to this study were identified through the first survey step—participatory community assessments. With the help of community leaders and extension workers, a total of 500 households from eleven villages from the Meghraj watershed were randomly selected and interviewed, representing 21%–24% of total village households. Similarly, a total of 300 households were interviewed from five villages in the Dharta watershed, representing 24%–29% of the total village households. Interviewees were either household heads or members who make decisions on behalf of household members.

Social and economic data were collected using a pre-tested questionnaire. Four major aspects were considered: (i) Household's livelihood assets—human, natural, physical, financial and social assets [9]; (ii) household livelihood activities and strategies; (iii) household's perceptions of livelihood determinants, potential future changes, and adaptive intentions; and (iv) farming inputs and outputs. A pilot survey in both watersheds was carried out to finalize the questionnaire before full-scale surveys were conducted.

A separate survey was also conducted to answer research question about women's responsibilities regarding water and gendered perceptions of water use, availability and quality and who collects water. Five villages from Gujarat and Rajasthan were chosen and an average of 10 women, three men and three members of community associations were interviewed from each village. A random sampling method was used. Both surveys mentioned here were translated in Hindi and Gujarati and field investigators underwent a three-day training session conducted by the MARVI research team.

Cluster Analysis was used to identify relatively homogeneous groups of households/farmers based on selected groundwater use characteristics. Because the goal of this cluster analysis is to identify a typology of similar groups of groundwater users, the agglomerative hierarchical clustering method was used in this study.

Engagement with Schools and Local Communities

Engagement of village communities through a range of activities that involved farmers, school communities and other members of village communities was an important part of the transdisciplinary approach used in this project. Field demonstrations on farmers' fields in the middle and end of the crop seasons were conducted on aspects, such as water requirements and water conservation practices, such as mulching and crop varieties that may be more drought tolerant or may result in improved income for a given water use. School children and teachers were engaged to record daily rainfall. Total weekly, monthly and seasonal values of rainfall were displayed on school noticeboards by students to create awareness about rainfall patterns and amounts and general awareness about water issues in their local areas.

Photovoice workshops were organized in villages and schools in both watersheds. Students and farmers were trained in photography and interestingly, most of them had never touched a camera in their lives. The idea of using a camera to express their ideas was something new and exciting for them and they actively participated in these workshops. They captured photographs regarding their past, present and future thoughts about water resources and groundwater as one of the critical factors of livelihood in village communities.

A newsletter in Hindi, called "MARVI Manthan"—a Hindi word Manthan meaning "deep contemplation"—was launched to share the project findings with village communities. This newsletter is published twice a year to coincide with the beginning of *Rabi* and *Kharif* seasons. The target audiences of this newsletter are farmers, the general community and other stakeholders, and the main purpose of the newsletter is to connect with local communities and pursue a dialogue with farmers for participatory use and management of local groundwater resources.

RESULTS

Understanding the Local Groundwater Situation

The water table fluctuation for the Dharta and Meghraj watersheds, based on the monitoring of the Central Ground Water Board of India, are shown in Figure 5 and Figure 6 respectively. The water table depth trends for the Dharta watershed indicate a depleting of groundwater from January 1994 to November 2000, after which the groundwater level seems to stabilize over the next 14 year period. From the twenty-year analysis, the net rate of groundwater depletion is of the order of 0.18 m per year. However, from the 2005 to 2013 period, the groundwater levels are increasing at the rate of 0.36 m per year. Over the 20-year period, the largest water table fluctuation, *i.e.*, the maximum difference between the lowest and highest groundwater level, was estimated to be 24 m. For the Meghraj watershed, the 18-year time series of groundwater levels show some large fluctuations in the water table but overall the net depletion in groundwater over 18 years seems to be negligible (Figure 6). The largest water table fluctuation for the Meghraj watershed was 8 m for the monitoring period considered.

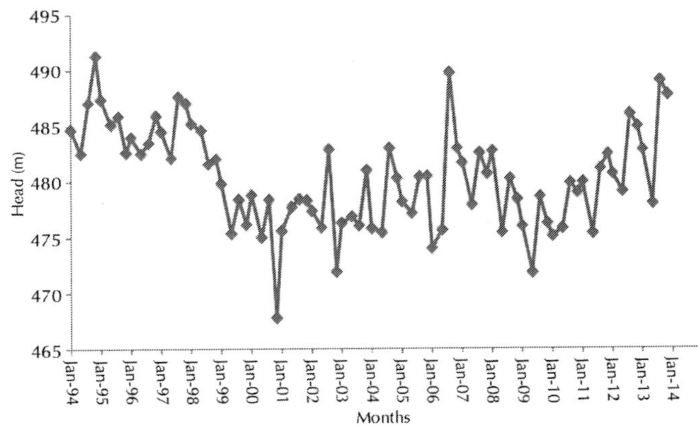

Figure 5. Central Ground Water Board groundwater head trends for the Dharta watershed from January 1994 to November 2013.

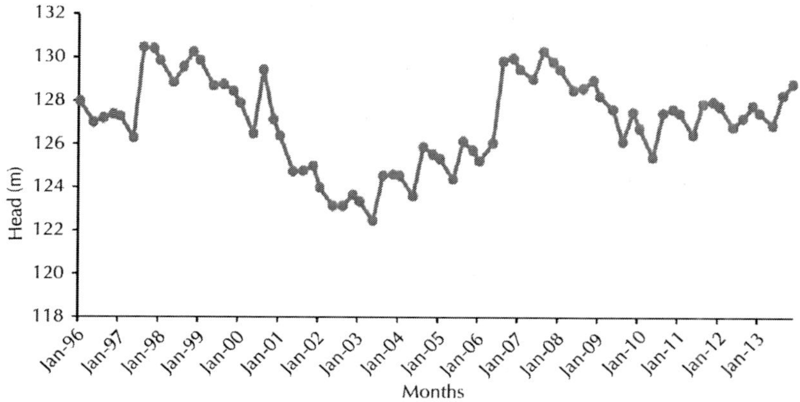

Figure 6. Central Ground Water Board groundwater head trends for the Meghraj watershed from January 2005 to November 2013.

Bhujal Jankaars

While the BJs were monitoring the water table on a weekly basis, they have also helped to develop good linkages between this project and local communities, creating awareness about the groundwater issues in the two watersheds. The evaluation of the BJ approach so far indicated that BJs interact extensively with their communities as they do their measurement tasks on a weekly basis. They are sharing project outputs that are written in the local language and tailored to the needs of village communities, particularly sharing some observed water table data to indicate the state of groundwater fluctuations in the area.

Discussion with village communities indicates that BJs are now becoming an integral part of the engagement process and data collection activities in the project in both watersheds. There is now an increasing acceptance of BJs in village communities in regards to the source of information about the local rainfall, extent of water table fluctuations and groundwater quality. They have also become an important link between the project team and the village communities for mobilizing farmers for project meetings, field demonstrations and dissemination of research findings from the project.

Engaging with Local Community

The engagement activities with the farming community through water table monitoring, crop demonstration, work with local schools and targeted workshops have helped to create community awareness about the local groundwater situation and develop a suitable atmosphere for future meaningful dialogue with the community on local groundwater management issues and challenges. Competencies are being built to enhance the social capital of the area, with the aim of facilitating mutually beneficial collective action. The workshops with villagers have indicated that farming and village communities were all deeply concerned about groundwater quality and the rapidly declining groundwater supplies. The community is willing to explore options that will help in improved water availability for irrigation and drinking purposes but are currently more focused on water availability in their individual wells and not appreciating that groundwater needs to be managed at the village and watershed levels and beyond.

The school engagement activities in the project included a poster and painting competition on a range of topics such as drip irrigation, water harvesting, soil testing and climate change. It was observed that the engagement of school children in the project extended the groundwater dialogue with parents and may also result in longer-term benefits. Another important engagement activity was photovoice workshops that involved school communities and villagers and resulted in several hundred photos and the subsequent selection of over 50 photos with text from the participants. Photovoice is essentially a participatory process of collecting information and expressing issues and concerns through photographs, and it can be used to effectively engage different groups and communities in a research project. It can particularly help individuals and communities groups to record and reflect on their ideas and concerns, help them promote critical dialogue and exchange of knowledge about important issues at different levels and reach policymakers for improving situations. Photovoice in this study helped to significantly engage teachers, students and villagers and facilitated them to think about their current groundwater situation and some options they may like to pursue to improve the situation. In particular, through a participatory photography process, the activity helped to explore some basic questions regarding what water means to villagers in the two watersheds. The analysis of photographs and text provided by the participants indicated that women and youth tend to emphasize future and personal responsibility while older male participants focused more on current problems. Unsurprising, the photovoice data indicated that participants saw the lack of water as the overarching problem, alongside specific human behavior and infrastructure problems.

In general, engagement and awareness campaigns aimed at educating the beneficiaries on a potential policy may be more effective, rather than using uninformed preferences based on expert opinion to drive policy decisions for complex natural resources management issues and challenges [10] (Rogers, 2013). In this project, the engagement with the local community and stakeholders has been an important element of transdisciplinary research on complex issue such as groundwater sustainability and will particularly assist in an effective dialogue with village communities, government agencies, including policy makers at the state and national levels, for participatory management of groundwater.

Socio-Economic Dimension of Groundwater Management

A series of 11 questions in the livelihood survey elicited household attitudes and perceptions concerning the role of MAR, adequacy of groundwater to meet future needs, the influence of extraction of, and on, proximate wells, mechanisms to coordinate aquifer resources, who should pay or be compensated for aquifer remediation and willingness to adjust extraction for future needs.

In this study, it was assumed that the cluster analysis can be used to identify relatively homogeneous groups of households/farmers based on selected groundwater use characteristics. There are numerous ways in which clusters can be formed and the hierarchical clustering is one of the most straightforward methods to use. Hierarchical clustering can be either agglomerative or divisive. Agglomerative hierarchical clustering begins with every case being a cluster unto itself. At successive steps, similar clusters are merged. The algorithm ends with everybody in one huge, but useless, cluster. A divisive clustering starts with one large cluster with all objects in it and gradually broken into smaller sized clusters and ends up with clusters with one object (singleton cluster).

Because the goal of this cluster analysis is to form similar groups of groundwater users, the agglomerative hierarchical clustering method is used in this study.

The cluster analysis of the factor scores in this study revealed a four-cluster solution (Table 1). Cluster composition and membership was predicted by eleven groundwater questions specified as x-axis variables. The composition and relative values of the four groundwater management clusters mainly differentiated attitudes regarding the effectiveness of MAR, the willingness to reduce extraction for their children's future use, the role of markets in groundwater management and relative impacts of proximate wells. The four clusters were defined:

- Cluster A-future and market oriented, with a preference for MAR;
- Cluster B-future, non-market oriented with a focus on water use efficiency;
- Cluster C-present non-market orientation; and
- Cluster D-present market orientation.

The present, markets groundwater management (Cluster D) is characterized by a low likelihood of children taking over the farm in the future, does not believe that increasing the depth of the well will have an impact on neighbors, does not consider that MAR is the best way to maintain the well, does not deem that efficient water use is the best way to maintain the well but expects that a MAR scheme operated by a neighbor and self should be compensated. In contrast, the future, markets, MAR groundwater management (Cluster A) is typified by a high likelihood of children taking over the farm, the belief that increasing the depth of the well had an impact on neighbors, judge MAR as the best way to maintain groundwater resources, and believe that a neighbor's groundwater use reduced water in their own well.

Table 1. Response to different questions for different clusters

Groundwater Attitudinal Questions (Yes/No Response)	Cluster (Proportion Yes Response)			
	GW A	GW B	GW C	GW D
How likely is it that your children will take over your farm in the future?	0.76	0.82	0.58	0.46
Do you think that increasing the depth of your well has had an impact on your neighbours?	0.78	0.51	0.08	0.00

Will the current depth of well/ tubewell be sufficient in the next 5 years for your current cropping pattern?	0.21	0.13	0.11	0.25
Is MAR the best way to maintain your well?	0.76	0.30	0.37	0.01
Is efficient water use the best way to maintain your well?	0.86	0.91	0.50	0.16
Has your neighbour's groundwater use reduced the amount of water in your well?	0.89	0.93	0.04	0.15
Would you be willing to share the water and costs of a recharge scheme with other farmers close to you?	0.96	0.73	0.32	0.86
Would you be willing to reduce the number of watering if it meant that water would be assured for your children?	0.92	0.88	0.44	0.30
If your managed recharge scheme increases the water available for your neighbours, should they compensate you?	0.99	0.08	0.05	0.93
If your neighbours managed recharge scheme increases the water in your well, should you pay them?	0.99	0.11	0.07	0.96
Would you be willing to adopt a new groundwater management scheme that shared water and costs fairly amongst all irrigators in your village?	1.00	0.99	0.82	1.00

The relative proportions of groundwater management cluster membership of respondents located in the two watersheds vary. About 9% of respondents from Gujarat are assigned membership in Cluster D, and 34% in Cluster A. The Rajasthan respondents are characterized by high proportional membership in Cluster D (55%) and Cluster A (40%). The farmers in Cluster D derive their groundwater information from traditional knowledge (42%), family (20%) and neighbors (19%), while those in Clusters A and B acquire information from family, neighbors and television. However, the farmers in Cluster C only rely on traditional knowledge (26%) and family (22%). As to the level of trust, there is no significant difference among the four clusters.

The cluster analysis indicates that groundwater management perceptions and attitudes influence the willingness and capacity of well owners to adopt specific remediating technological solutions and their compliance with policy incentives. Differentiated perceptions and information sources revealed in the cluster membership and the distribution of clusters in the two watersheds suggests that a suite of targeted technologies and incentives, in contrast to a reliance on single technological solutions and policy instruments, is likely to achieve the highest adoption rates [11]. The analysis provides the basis for designing watershed specific policy instruments and technologies that align with statistically differentiated attitudes and perceptions revealed in the four clusters.

Groundwater and Gender

Though women are found to be significantly involved in irrigated agriculture in both the Dharta and Meghraj watersheds, the revenue generated from agriculture is entirely controlled by men. This clearly separated intra-household activities according to gender. These activities, however, are not separate from the water users' perspective, and this often impedes women's access to and control over this scarce resource. For instance, men usually have a greater say in water provision for irrigated agricultural production, which in turn influences agencies responsible for infrastructure and determines availability and security of water from the women's perspective. Even production from women's fields and household gardens is often controlled by men to a certain degree, as is the availability of water for non-agricultural tasks. This bias of water allocation and control is even greater in times of water scarcity.

Women were found to be responsible not only for domestic water use but also in the productive uses of water, such as vegetable growing and herding. The women interviewed are almost exclusively responsible for domestic chores and for maintaining hygiene in their households. Most of them commented that water scarcity has a direct impact on their access to water within the household as well as on the time they and their daughters and daughters-in-law have to spend in water collection. This means the time available for other activities in the household and livelihood opportunities becomes limited. In addition, mothers are concerned that their daughters are missing school because they have to help in water collection. A majority of them suggested boosting women representation in groundwater management.

The women interviewed are almost exclusively responsible for domestic chores and for maintaining hygiene in their households. Due to inadequate water being locally available for basic consumption in poorer households, women fetch water from nearby villages, where applicable, walking for more than 30 min and up to one hour per trip. The physical strain of collecting water is doubly compounded during the peak of summer, and women have to wait in long queues at water sources. This shows the precarious situation of women in households and also indicates how women are compelled to shoulder additional burdens for the welfare of their families.

Overall, the analysis of gender related issues of water indicate that for achieving broad livelihood improvement outcomes the needs of water from women's perspective cannot be ignored. Furthermore, the gender aspect of groundwater needs to be considered along with securing sustaining groundwater for crop production.

DISCUSSION

Capacity Building of BJs as Local Champions

The training program of eight modules spread over about six months was aimed to orient the BJs regarding the MARVI project and to build their understanding about geology, hydrology, watershed management and mapping. While it was comparatively easy to develop an understanding of the depletion of surface water resources, the measures used for water harvesting and groundwater issues are quite complex to comprehend, both for the village communities as well as the project research partner field staff. However, in presenting these module inputs it was realized that, despite the difficulties, it was possible to demystify the technical aspects of groundwater management in a language that villagers could understand. It was also recognized that capacity building for the BJs has to be a gradual and continuous process, and one which blends theoretical inputs with practical exercises in their own villages in order to help them grasp these complex issues. Convincing people to work on an action research project that does not give them direct benefits requires a lot of effort. In addition, it was observed that groundwater management is a new concept that is not easily understood by rural communities. Retaining the BJs in the midst of other work opportunities available in and near the villages at a high remuneration was also an issue.

Besides well monitoring, the BJ's were also linked into other key project activities, viz.; village level meetings, field days during and after crop demonstrations and seed and fertilizer distribution. The BJs shared their experiences in a monthly meeting with project community organizers and prepared the plans and strategies for further activities. The BJs also interacted with other village members individually or through various village institutions like farmers' club, *Sujal Samiti* (water co-operative), *Gramsabha* (village council) and the like. In this way, the BJs were working as a communication bridge between the MARVI project team and villagers.

An interesting aspect of BJ involvement in this project was that the information collected by BJs made people in the villages curious about the MARVI project activities and triggered further communication. The location of monitoring wells also helped in spreading information as the wells were widely dispersed and every well owner asked why was the BJ taking readings and what will come out of it? These questions assisted in starting communication with the farmers about the current issues of groundwater scarcity. Some of the BJs became quite capable in preparing charts for displaying current rainfall and well water depth and hung these outside their house so that more people could see the results. Thus, as result of the BJ's involvement in the project most people in the villages came to know that this is a research project, not another project that focus on on-ground construction works, and that the research data which are being collected will be helpful for them in the future.

Given the skills that the BJs have acquired through their training, and subsequent practical experiences, they will be able to continue to contribute towards various development projects being implemented by local NGOs and the Government agencies, both in their own and the adjoining villages. It was observed that there was a dearth of competent human resources available at the village level before the commencement of this MARVI project, but now local villagers have relevant local groundwater knowledge, data collection experience and significant interest to improve their groundwater situation. It would be foolhardy for anybody not to utilize the local knowledge and skills on water and agriculture acquired by these BJs. At least one project partner is now collaborating with the Government of Gujarat to promote and effectively use the BJs to help implement the Mahatma Gandhi National Rural Employment Guarantee Act (MGNREGA), while in Rajasthan the

BJs can continue to work with other government funded watershed development projects that are about to commence in the watershed. This is also in line with a recent report by the Planning Commission of India highlighting the need for building strong partnerships and collaborations among a broad spectrum of institutions and community to monitor and implement groundwater management strategies across India [12].

Managing Complexity of Groundwater Use

Farmers in the two watersheds face significant water shortages and the risk of crop failure even with a slightly abnormal decline or delay in monsoonal rains. Because of advances in drilling technology and its easy access, there has been a massive increase in the drilling of tubewells and deepening of open wells for irrigation. This has motivated farmers to extract groundwater from whatever depth it is available. As a consequence, this phenomenon has changed the idea of equity and sustainability of groundwater use in the two watersheds. Not only is the water table lowering or fluctuating considerably from year to year, which impacts on crop production but, also the quality of groundwater has deteriorated due to pumping from deeper aquifers. For example, in the Dharta watershed, there is some evidence that fluoride levels in groundwater (which is also used for drinking water supplies) for some villages are above the values recommended in the World Health Organization's guidelines [13]. In general, the groundwater situation in the two study watersheds also illustrates what is prevailing in many other parts of the States of Gujarat and Rajasthan and for that matter in many States of India.

Another complex and difficult issue is determining the limits of groundwater available for withdrawal, especially in hard rock aquifers with limited storage capacity. Without mechanisms and sanctions to coordinate individual withdrawals to meet socially agreed sustainable levels, groundwater use represents an open access resource where at the end everyone loses when the groundwater system gets over exploited. Access to and availability of groundwater affects household livelihoods and community well-being and, in some instances in India, it has been reported to have led to farmers taking the extreme step of ending their own lives [14]. Therefore, a proposal to coordinate groundwater use remains a source of conflict between competing farmer interests and is the subject of significant political argument. The flow of groundwater does not recognize boundaries of individual farms, villages or watersheds and the subtractive attribute implies that one farmer's gain through over-pumping incurs a loss of access for others. Therefore, in the current situation it is almost impossible to ensure equity of access among farmers and regulate its use sustainably.

While groundwater recharge of varying amounts occurs during each monsoon season, there has been a net lowering of water tables in many parts of Gujarat and Rajasthan [4]. The consequences are notably manifest during the *Rabi* season. In the absence of institutions, regulations to share the costs and risks of aquifer remediation, individual farmers are unlikely to undertake mitigating actions independently, as they are unlikely to be compensated for the benefits shared by the common pool community. The depth to the water table increases with pumping over a longer time period, and the impact of such pumping usually extends over larger areas. While groundwater recharge of varying amounts occurs during each monsoon season, the real impact of any lowering of the water table is severely felt during drought periods. Once groundwater has been extracted in excess of annual recharge, it is not easy for individual farmers to reverse this situation. It then requires co-operative actions from group of adjoining farmers to see any real impact of local recharge and demand management on the water table situation.

Challenges of Sharing Groundwater

It is important to recognize that groundwater is an invisible, common property resource that is accessible to anyone who has a well and a pump, or can afford to dig a well and install a pump. The amount of groundwater available in hard rock aquifers with their limited storage capacity is not easy to predict, and hence it is hard to estimate the limit of groundwater pumping. Groundwater use is a good example of "tragedy of the commons" and "survival of the fittest" but at the end everyone loses when the groundwater system is over exploited. Groundwater can affect the livelihood and wellbeing of communities. Therefore, the regulation of groundwater use is a very sensitive issue for farmers and can become a significant political issue if not tackled properly

Common pool resources are characterized by costly exclusion of beneficiaries, a characteristic shared with public goods and rival consumption (or subtractable usage), a characteristic shared with private goods [15]. That is, the withdrawal of additional groundwater by an individual well owner appropriates and subtracts from the total available aquifer volume, reducing the opportunity of other irrigators to make use of the groundwater resource. When joint outcomes depend on multiple actors contributing inputs or actions that are costly and difficult to quantify and there is a lack of policy instruments to restrict usage, incentives exist for individuals to act opportunistically, often appropriating to a level where aggregate overuse occurs. A social dilemma occurs when individuals are tempted by short-term gains to over appropriate the common pool resource, thereby imposing group-shared costs on the common pool community. Additionally the opportunity exists for some individuals to free ride and benefit from the reductions in extraction or increases in recharge committed by other aquifer users. Individual over appropriation will eventually lead to falling water tables, increased pumping costs and lower crop productivity for all farmers.

The solution to the overexploitation of groundwater may well come from adequate licensing to access the resource. In India, the electricity for groundwater pumping is free in a number of states, and as such this has aggravated the problem of overuse groundwater. On the other hand, the State Government of Gujarat in recent years implemented a policy to limit groundwater pumping through limiting hours of electricity supply by constructing a separate power grid for farm sector. While the policy implementation in Gujarat has certainly limited the hours of pumping, this also pointed out that any attempt to deal with the issue of limiting user access to groundwater, in this case limiting the supply of electricity through a separate power grid, does involve some transaction cost of policy. An important outcome of the transdisciplinary research in this study would be to understand the issues and options of groundwater overexploitation from a number of perspectives and design a system of effective control for groundwater access.

Making Community Engagement Effective

Groundwater, being a common resource accessible by every member of the community individually, requires a common approach to its management. However, in general, past efforts of community involvement in aquifer management have been shown to be quite inefficient [4]. Therefore, for this study, it was decided to tackle this issue through more effective participation by the village communities involved, and thus community engagement was critically important to the success of the study.

Effective participation is important groundwater management and in general it depends upon commitment rather than coercion and cannot be fully programmed or tightly controlled. Further, it involves resolving issues about the nature of participation in terms of extent and quality, as well as questions about who should participate. Sriskandarajah *et al.* [16] identified key themes in participative projects and included (i) the importance of the scope for genuine participation in decision-making if "community participation" is to be meaningful; (ii) the need to see participation as a continuing process of negotiation and decision-making rather than a once only input into project planning; (iii) the need for clear identification of interested parties as the first step in establishing community based resource management programs; and (iv) the need to recognize and build upon local knowledge and existing local resource management and institutional support practices.

A number of different forms of "Citizen participation" have been identified by Arnstein [17] in the form of a ladder, which moves from very tokenistic forms of participation (manipulation) and progresses to more meaningful forms of involvement (Citizen control), as illustrated in Figure 7. In the context of resource management projects, Sriskandarajah *et al.* [16] also suggested that at the three higher levels, community participation involved local people in making decisions about the management of the resources they used, while at the lower five levels, these decisions were made by bureaucratic "experts", with community members only being involved as either voluntary or paid labor. At the higher order, participation meant that communities either defined the ends themselves, or at least had a substantial input in defining them.

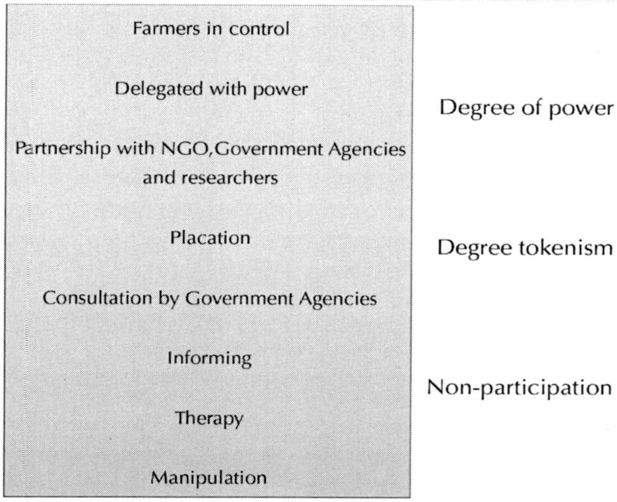

Figure 7. Degree of participation for managing groundwater (adapted from Arnstein, [17]).

Pretty [18] suggested that two overlapping schools of thought and practice have evolved. One views participation as a means to increase efficiency, with the central notion that when people are involved, they are more likely to agree with and support the new development or service. The other view sees participation as a fundamental right, in which the main aim is to initiate mobilization for collective action, empowerment and institution building. In an analysis somewhat similar to that of Arnstein [17], Pretty [18] notes that participation has been used to justify the extension of and control by the state, as well as to build local capacity and self-reliance; it has been used to justify external decisions, as well as to devolve power and decision-making away from external agencies; and it has been used for data collection, as well as for interactive analysis.

In this study we felt the problems Shah [4] had identified were due to participation being at level A, while we would use community engagement to strive to achieve level (B), but also developing local capacity to move to level C (Table 2). It is considered that we have achieved level B participation and that this is continuing to strengthen as the project matures. As research results become available and are shared through the community engagement processes, notably via the

BJs, it is hoped that the options for improvement to ensure groundwater sustainability will be taken up and lead to the emergence of level C participation.

Table 2. A selection of the typology of participation: How people participate in development programs and projects (adapted from Pretty [18])

Participation Level	Participation Type	Description
A	Functional Participation	Participation seen by external agencies as a means to achieve project goals, especially reduced costs. People may participate by forming groups to meet predetermined objectives related to the project. Such involvement may be interactive and involve shared decision-making, but tends to arise only after major decisions have already been made by external agents. At worst, local people may still only be co-opted to serve external goals.
B	Interactive Participation	People participate in joint analysis, development of action plans and formation or strengthening of local institutions. Participation is seen as a right, not just the means to achieve project goals. The process involves interdisciplinary methodologies that seek multiple perspectives and make use of systemic and structured learning processes. As groups take control over local decisions and determine how available resources are used, so they have a stake in maintaining structures or practices.
C	Self-Mobilization	People participate by taking initiatives independently of external institutions to change systems. They develop contacts with external institutions for resources and technical advice they need, but retain control over how resources are used. Self-mobilization can spread if governments and NGOs provide an enabling framework of support. Such self-initiated mobilization may or may not challenge existing distributions of wealth and power.

CONCLUDING REMARKS

Sustainable groundwater use is a wicked problem and has technical, social, economic, policy and political dimensions. The access to groundwater for the farming communities is also an emotional issue as their livelihood and survival depends on it. Availability of relevant and reliable data related to the various aspects of groundwater and developing trust and support between local communities, NGOs and government agencies are the key to moving towards a dialogue to decide on what to do to achieve sustainable use of groundwater. Technical information on water table fluctuations, groundwater balance modeling, socio-economic and other data and analyses alone will hardly have any impact on over-exploitation of groundwater resources. This study has demonstrated that transdisciplinary research, which involves people who are going to benefit, is more effective in developing a deeper understanding of issues and exploring options to improve the current groundwater situation. In particular, the involvement of local villagers through groundwater monitoring, photovoice techniques and community workshops has been valuable in generating local knowledge and capacity building.

The socio-economic analysis revealed diverse attitudes to farmers' own and neighbors' groundwater responsibilities, mechanisms to coordinate groundwater use, attitudes to MAR, information sources and preferred groundwater and MAR managing agencies. Cluster membership variance highlights three key factors in designing participatory approaches and potential ground water management instruments in the two study sites. First, design principles need to address the diversity of attitudes and motivations observed in the sampled households, by emphasizing the participation of members across the whole cluster typology. Second, a reliance on a single instrument or approach to coordinate aquifer access is unlikely to align with the diverse attitudes observed across clusters, potentially resulting in low compliance rates or antagonizing sustainable groundwater management and MAR efforts. Third, while the transaction costs and resource demands make the tailoring of instruments to correspond with cluster attributes infeasible, community consultation is likely to reveal instrument sequencing as a viable strategy to promote aquifer sustainability. Addressing these three design principles in response to the observed household diversity is likely to enhance the prospects of community participation and improve aquifer recharge and groundwater pumping coordination.

The project has demonstrated that the harnessing of local experience and the indigenous knowledge of villagers has been useful in understanding the real issues of groundwater management, the geology of the area and groundwater use and changes over time. This engagement also helped in creating awareness about the project and sensitizing the community about the concept of groundwater management. The community is well aware that their groundwater is depleting at a fast rate but they were not aware of the technical reasons behind it. Local villagers had the perception that by digging deeper tubewells they would have more water, but they were not examining the issues related to groundwater recharge and water quality management. The regular monitoring of wells by BJs and the subsequent community meetings and the presence of project staff in the two study areas has now prompted the communities to talk among themselves about the future of their groundwater resources and the need to find options for managing and using groundwater more sustainably.

Efforts have been made by various government and NGO's for the augmentation of the water table, but this has not been enough to ensure long term sustainability. There is a need to awaken the people to take up groundwater recharge and rainwater harvesting and also manage demand and make irrigation more efficient. This is where the local administration must take responsibility and ensure that villagers are fully involved in such schemes. Planning is required at the micro level using participatory approaches to make each village self-sufficient in water.

ACKNOWLEDGMENTS

Funding for this research was provided by the Australian Centre for International Agricultural Research, Canberra, Australia, and we appreciate the support of Evan Christen, Research Program Manager, Land and Water Resources during this study. Thanks to Joycelyn Applebee, Sita Ram Bhakar, Pradeep Bhatnagar and Anand Singh Jodha for their assistance during the study.

REFERENCES

1. World Bank. *Deep Well and Prudence: Towards Pragmatic Action for Addressing Groundwater Overexploitation in India*; The World Bank: Washington, DC, USA, 2010; p. 97.

2. Siebert, S.; Burke, J.; Faures, J.M.; Frenken, K.; Hoogeveen, J.; Döll, P.; Portmann, F.T. Groundwater use for irrigation—A global inventory. *Hydrol. Earth Syst. Sci.* 2010, *14*, 1863–1880.
3. Shah, T. India's groundwater irrigation economy: The Challenge of balancing livelihoods and environment. In *Handbook on Environmental Economics in India*; Chopra, K., Dayal, V., Eds.; Oxford University Press: New Delhi, India, 2008.
4. Shah, T. *Taming the Anarchy: Groundwater Governance in South Asia*; Routledge: New Delhi, India, 2009.
5. MNREGA (Mahatma Gandhi National Rural Employment Guarantee Act). Available online: http://www.nrega.ap.gov.in/ (accessed on 5 September 2014).
6. Uphoff, N.; Wijayaratna, C.M. Demonstrated Benefits from Social Capital: The Productivity of Farmer Organisations in Gal Oya, Sri Lanka. *World Dev.* 2000, *28*, 1875–1890.
7. Central Ground Water Board (CGWB). *Master Plan for Artificial Recharge to Ground Water in India*; Government of India, Ministry of Water Resources: New Delhi, India, 2013. Available online: http://cgwb.gov.in/documents/MasterPlan-2013.pdf (accessed on 6 November 2014).
8. India-WRIS. Water Resources Information System of India. Available online: www.india-wris.nrsc.gov.in (accessed on 7 November 2014).
9. Chambers, R.; Conway, G.R. *Sustainable Rural Livelihoods: Practical Concepts for the 21st Century*; IDS Discussion Paper No. 296. Institute of Development Studies: Brighton, UK, 1991; p. 29.
10. Rogers, A.A. Public and expert preference divergence: Evidence from a choice experiment of marine reserves in Australia. *Land Econ.* 2013, *89*, 346–370.
11. Ward, J.; Dillon, P. Design principles to coordinate managed aquifer recharge with natural resource management policies in Australia. *Hydrogeol. J.* 2012, *20*, 943–956.
12. Planning Commission of India. *Faster, Sustainable and More Inclusive Growth: An Approach Paper to 12th Five-Year Plan (2012–2017)*; Planning Commission, Government of India: New Delhi, India, 2011; p. 137.

13. Mishra, P.; SWATCH, Udaipur, Rajasthan, India. Personal communication, 2014.
14. Govardhan Das, S.V.; Rao, P.S.; Mani, K.A.S. Practices that can combat Poverty and Distress in India. In Proceedings of the 128th Session of International Conference on Agrarian Reform and Rural Development (ICARRD): New Challenges and Options for Revitalising Rural Communities, Porto Alegre, Rio Grande de Sul, Brazil, 7–10 March 2006; p. 12.
15. Ostrom, E. How types of goods and property rights jointly affect collective action. *J. Theor. Polit.* 2003, *15*.
16. Sriskandarajah, N.; Fisher, R.J.; Packham, R.G. Community Participation in Natural Resource Management: Lessons from Field Experience. In Proceedings of the Regional Seminar "Ecotone V" on Community Participation in Conservation, Sustainable Use and Rehabilitation of Mangroves in Southeast Asia, Ho Chi Minh City, Vietnam, 8–12 January 1996.
17. Arnstein, S.A. A Ladder of Citizen Participation. *J. Am. Inst. Plan.* 1969, *35*, 216–224.
18. Pretty, J. Participatory Learning for Integrated Farming. *Internet Conference on Integrated Bio-Systems*; Institute of Advanced Studies, UN University: Tokyo, Japan, 1998. Available online: http://www.ias.unu.edu/proceedings/icibs (accessed on 7 November 2014).

Chapter 8

Assessing the Feasibility of Managed Aquifer Recharge for Irrigation under Uncertainty

Muhammad Arshad [1], Joseph H.A. Guillaume [1, 2] and Andrew Ross [1,3]

[1]National Centre for Groundwater Research and Training and The Fenner School of Environment and Society, Integrated Catchment Assessment and Management Centre (iCAM), building 48a, Linnaeus Way, The Australian National University, Canberra, ACT 0200, Australia

[2]Water & Development Research Group (WDRG), Aalto University, Tietotie 1E, Espoo 02150, Finland

[3]Groundwater Section, Division of Water Sciences, UNESCO, 1 Rue Miollis, Paris, Cedex 75732, France

ABSTRACT

Additional storage of water is a potential option to meet future water supply goals. Financial comparisons are needed to improve decision making about whether to store water in surface reservoirs or below ground, using managed aquifer recharge (MAR). In some places, the results of cost-benefit analysis show that MAR is financially superior to surface storage. However, uncertainty often exists as to whether MAR systems will remain operationally effective and profitable in the future, because the profitability of MAR is dependent on many uncertain technical and financial variables. This paper introduces a method to assess the financial feasibility of MAR under uncertainty. We assess such uncertainties by identification of cross-over points in break-even analysis. Cross-over points are the thresholds where MAR and surface storage have equal financial returns. Such thresholds can be interpreted as a set of minimum requirements beyond which an investment in MAR may no longer be worthwhile. Checking that these thresholds are satisfied can improve confidence in decision making. Our suggested approach can also be used to identify areas that may not be suitable for MAR, thereby avoiding expensive hydrogeological and geophysical investigations.

INTRODUCTION

Water demand continues to grow in order to maintain food security and drinking water supplies, while supplies remain limited from conventional sources. Future water security is threatened in many places, as most suitable locations for large surface storages have already been used [1] and ground water is often being withdrawn at unsustainable rates [2,3,4]. Among other options of water supply augmentation, such as water recycling, desalination *etc.*, storing more water underground appears to be a potential solution to achieve future water supply goals. For many water stressed areas, water security and reliability do not necessarily depend on the absolute amount of precipitation, but on the fraction of water that is efficiently retained as storage for future use [5].

Water shortages can be eased by storing surplus water underground during wet periods for later use during dry periods. Managed aquifer recharge (MAR) has been used successfully in several countries for the storage and treatment of water [6,7,8,9]. Storage of surplus water in aquifers can help minimize evaporative losses and help irrigators to adjust to surface water variability during droughts, provided that MAR is technically feasible and cost effective. The feasibility of MAR and its comparative cost to other alternatives depend on a number of technical and financial factors, such as infiltration, injection and recovery rates, which are dependent on local hydrogeology [10].

A few studies indicate that MAR can achieve more financial value than surface storage and other alternatives [11, 12]. However, uncertainty often exists whether it is more cost effective to store water above ground in surface reservoirs or below ground using managed aquifer recharge [13].

Cost-benefit analysis (CBA) provides a comparison of benefits and costs resulting from a proposed policy or investment [14]. Previous studies undertaking CBA of MAR have assumed hydrogeological factors, such as infiltration, injection and recovery rates, to be known [11, 12, 15]. Overlooking such uncertainties can result in lower than expected operational efficiency and irrigation returns from MAR [16,17]. For example, future returns from MAR may be affected by increases in groundwater pumping cost or reductions in infiltration rates.

An increase in the turbidity of source water due to hydrological variability can significantly increase the cost of infiltration basin maintenance, adding to the cost of water quality treatment for aquifer storage and recovery (ASR) systems. Maliva [16], in this special issue, highlights that assessing such uncertainty is perhaps the most neglected aspect in the economics of MAR.

The primary focus of this paper is to systematically search for conditions under which the requirements for MAR may not be met and failure might occur. Playing such a devil's advocate role has been shown to improve decision making compared to an exclusively expert-driven approach [18]. The approach used identifies thresholds above which MAR is financially better than surface storage and below which it is not. These thresholds (or cross-over points) describe corresponding values of variables at which the net present value (NPV) from MAR and surface storage become equal. All dollar amounts reported in this study are in Australian dollars. An example of a cross-over point for pumping cost is shown in **Figure 1**, where basin infiltration (red line) and surface storage (green line) options are compared; and where basin infiltration is initially (dashed vertical line) more profitable than surface storage. A cross-over point between the two compared options is possible when the cost of pumping increases from the best guess value of 35 $/megalitres (ML) to 53.63 $/ML. This increase in the pumping cost will decrease benefits (NPV) from basin infiltration, such that they become equal to the benefits (NPV) obtained from surface storage. However, aquifer storage and recovery (ASR) always result in an inferior NPV regardless of the pumping cost. There is no cross-over point between ASR and the other alternatives.

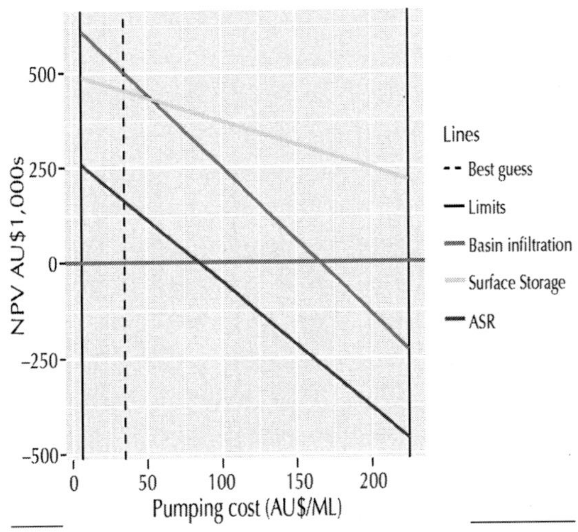

Figure 1. Illustration of identifying cross-over points for pumping cost when comparing basin infiltration, aquifer storage and recovery (ASR) and surface storage of irrigation water.

At the cross-over point, the decision maker is indifferent to choosing a single option from the two, because their financial returns are equal. In our method, we use computational techniques to identify the cross-over points as values of uncertain variables where the NPV of MAR is exactly equal to the NPV of surface storage of irrigation water. The approach is demonstrated through a case study in a highly developed irrigation region of the lower Namoi catchment in New

South Wales, Australia, where irrigation water restrictions motivate the need to consider options to supplement future irrigation supplies, such as MAR.

The suggested approach of identifying cross-over points is beneficial in three ways;

- It can determine minimum hydrogeological and cost requirements under which MAR can be worthwhile;
- It can improve confidence in decision making for MAR investment, by enabling the assessment of conditions that are unfavourable to MAR compared to surface water storage;
- It can substantially lower the cost of geophysical and hydrogeological investigations by targeting only areas that satisfy the minimum requirements, as MAR investigations and trials are shown to be time and resource expensive.

The next section provides an overview of the literature on the feasibility of MAR with a focus on the technical, financial and uncertainty considerations. Section 3 ("Methods") describes the model and tool used to explore cross-over points. In Section 4, an illustrative study in the lower Namoi catchment evaluates the irrigation-related costs and benefits of storing flood water in aquifers compared to surface storages. The analysis of cross-over points in Section 5 provides a discussion of how cross-over points are reached when only a single variable changes, as well as when many variables interact.

RELATED WORK: FEASIBILITY OF MANAGED AQUIFER RECHARGE

Assessing the feasibility of MAR requires the integration of many types of data and information from many disciplines (Figure 2). Although carrying out a comprehensive feasibility assessment is essential, the first step in establishing an MAR scheme requires assessing the feasibility of technical and financial factors, to provide a basis for other investigations to proceed.

An overview of the basic requirements and feasibility guidelines for managed aquifer recharge (MAR) is available in [10] and GHD and AGT [19].

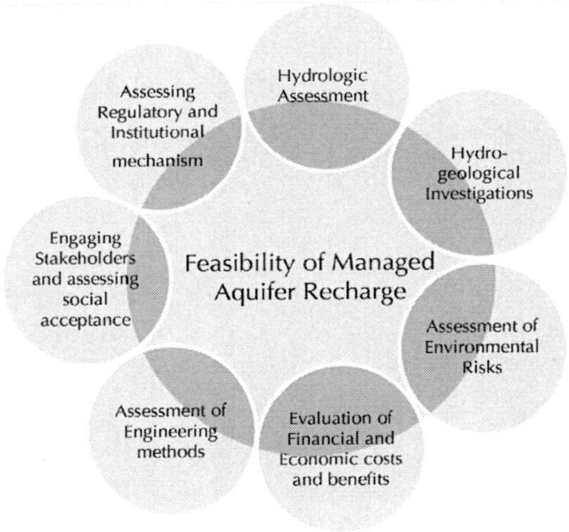

Figure 2. A framework for the feasibility of managed aquifer recharge: adapted from GHD and AGT [19], Dillon *et al.* [10] and Rawluk *et al.* [20].

Technical Considerations

Key technical requirements for MAR include hydrogeological assessment of the target aquifer, the availability of surplus surface water and the means to convey it underground. Relevant hydrogeological factors include aquifer storage size, permeability, infiltration, injection and recovery rates and connections with other aquifers [21,22]. High infiltration rates lower the cost of underground storage; for example, a basin infiltration system with high infiltration rates will require a smaller pond area and can be cheaper to construct and maintain than a pond with low infiltration rates.

There are two main types of MAR methods: basin infiltration and aquifer storage and recovery (ASR), each favourable to different hydrogeological conditions. Basin infiltration is suitable to recharge shallow unconfined aquifers with minimal or no treatment of the recharge water. The methods include deep, large diameter isolated wells, infiltration ponds, infiltration galleries, induced bank filtration, leaky and recharge dams and redirecting floodwaters over the wider landscape to supplement areal recharge [7,9]. Some basin infiltration methods require large surface areas and permeable soils to be effective [21, 23].

ASR involves the injection and recovery of water using wells; this has the advantage of targeting a desired aquifer for recharge. Thus, zones of saline water or clay layers can be bypassed. However, ASR systems are costly because of the need for bore well construction and water treatment prior to recharging, and if clogging occurs, they are costly to repair. Passive borehole recharge (under gravity) requires limited mechanical assistance, but the infiltration rate is relatively low. Water injection using pumps can greatly improve the rate of aquifer recharge [24,25]; however, the pumps require constant maintenance and are costly to run. The risk of clogging of the surface or well with fine sediments is common to both MAR methods. Solutions to this issue include stabilization of recharge water through settling ponds and treatment of water before recharge.

Financial Considerations

When the focus is on estimating the total economic benefits of recharge to a region instead of an individual, the benefits of aquifer storage become complex, as this needs to include public good, socio-economic and environmental benefits to a region, which are more difficult to assess and quantify. Maliva [16] in this special issue provides a greater review of the methods and techniques for assessing total benefits from MAR. With a known target volume of storage and recovery, it is easier to quantify the financial benefits, since the goal is the recovery of the stored water, and the volumes recovered accrue to an identifiable person or water utility for a particular use. The financial feasibility of MAR can then be studied in comparison to other water supply and storage alternatives, including surface storage.

The local situation dictates the costs of MAR options, and large variations may occur between localities [10]. For a fair comparison, it is essential to analyse the benefits and costs of MAR and surface storage in the same location, because the comparison of benefits and costs is complicated by the wide range of biophysical, socio-economic and regulatory conditions in which MAR occurs. There is little published analysis of the economic and financial benefits of MAR. From the few published studies, Ross and Arshad [26] compiled and reported the benefits and costs of surface storage and MAR at multiple locations, showing that the costs and returns of MAR options vary substantially.

Uncertainty Considerations

A number of methods have been used to address uncertainty in cost-benefit analysis. Sensitivity analysis simulates the impact of changes in financial behaviour, such as the change in NPV of an investment due to a change in an input variable, and identifies variables that are of greater concern [27]. Probabilistic analysis provides the combined effect of variables' variability on the financial behaviour [17]. Possibility theory assumes that all values within a certain range are possible, with the exact value being treated as unknown [27].

We focus on cross-over points as one possible means of addressing uncertainty in the cost-benefit analysis of MAR. Identification of cross-over points relies solely on the relationship between variables, such that it requires minimal understanding of the uncertainty of variables. The idea of a cross-over point is sufficiently simple that it has a number of widely used variations; it is also known as a break-even point or switch-over point. However, the term break-even in economics specifically applies to the volume of sales at which profit is zero as revenues cover total cost and is therefore used as a tool to calculate the margin of safety of a single investment [28], rather than comparing alternatives. The concept of a cross-over point is fairly simple with only one or two variables, but the complexity increases in the analysis and interpretation of results as the number of input variables increases [29].

METHODS

The analyses in this paper are carried out in two steps; in the first step, financial analysis compares the net present value of farm benefits to identify the best among the considered options. In the second step, the break-even analysis of cross-over points is carried out; this involves finding values of variables that will provide exactly the same financial returns from the two compared options. The variables were chosen based on an examination of literature concerning the financial feasibility of MAR. Identifying cross-over points allows the user to understand the minimum conditions required for success and allows measures to be taken to ensure they do not occur.

Financial analysis evaluates whether investment in MAR is worthwhile. Analyses of cross-over points help understand the circumstances when MAR is worthwhile. At the most basic level, MAR is worthwhile when net irrigation returns of MAR exceed those of alternatives. In our example, benefits are determined by the agricultural value of the additional water provided, by saving it from non-productive evaporation. This has been referred to as a "vapour shift" [30] from non-productive evaporation to agriculturally-valuable crop transpiration. Costs are composed of additional pumping to recover recharged water and MAR method-specific capital and ongoing costs of implementation during the life of the project.

To enable the break-even analysis, the financial analysis is programmed as a function in R [31]. As a general purpose statistical programming language, R offers a suite of optimization methods, as well as providing tools for visualization and the means to include a user interface. To identify cross-over points of single variables, other variables are set to fixed values, and the R function *uniroot* [32] is used to identify the value of the variable where the difference in NPV between the two compared options is zero (*i.e.*, $\Delta NPV(\Theta) = 0$), meaning that the two options have equal NPV. To identify cross-over points involving many variables, we use optimization to identify a cross-over point (*i.e.*, a point Θ where $\Delta NPV(\Theta) = 0$) that is closest to the best guess, in the sense of minimizing the maximum of the distances for each variable, expressed in relative terms using user-defined bounds ($\max_i |\Theta_i - \Theta_{best,i}| / |\Theta_{bound,i} - \Theta_{best,i}|$). This is one possible criterion for selecting cross-over points of concern. Other criteria, including probabilistic ones, would be possible and would usually raise different cross-over points for discussion. The code for the analysis is available online [33]. The cross-over points generated are assessed by comparing them to maximum and minimum values of variables that a decision maker thinks might be possible due to physical, climate or policy change over the analysis period. The resulting judgment of a cross-over point is not perfect and is based on the best available knowledge of the decision maker for each variable.

THE STUDY AREA: LOWER NAMOI

In many parts of Australia, overdraft of aquifers is resulting in falling groundwater levels in the shallow, unconfined systems and decreasing groundwater pressures in the deep confined and semi-confined systems [34]. In response to the groundwater overdraft, the New South Wales (NSW) government has reduced current groundwater entitlements in its stressed aquifer systems [35]. For the lower Namoi catchment, a highly developed cotton irrigation district in NSW, this cutback translates to a reduction of 21 gigalitres (GL)/year in groundwater entitlements for irrigation by 2015 and beyond. Groundwater in the Namoi River catchment supports an irrigation industry worth in excess of $380 million per annum [36]. All irrigation water is stored and routed from surface storages before application to the field. On-farm water storages within the lower Namoi range from conventional single-cell to advanced multi-cell farm dams. The typical Namoi valley farm holds enough water in storage to complete one full year of irrigation. Conservative estimates suggest that the total capacity of on-farm storages in the cotton industry could be on the order of 3150 gigalitres (GL). Evaporative losses from these surface storages are significant. On average, from surface water storages, evaporative losses range from 1200 to 1800 mm/year [37], which constitute 35% to 50% losses from surface water storage volumes.

To tackle the problem of reduced allocation and evaporative losses, improving water use efficiency at the farm level is an obvious option. This will include installing drip irrigation systems, lining water courses and further improving the design of surface storages to minimize evaporative and seepage losses. Improving water use efficiency needs to be a stepwise approach. Another potential option to reduce evaporative losses is to store water underground in aquifers using managed aquifer recharge. Recently, several studies have highlighted the potential of a regional-scale MAR project in the lower Namoi. Arshad *et al.* [38] indicated that a significant volume of water could be available from large floods for MAR while still satisfying environmental flow and ecological requirements. Similarly, Rawluk *et al.* [20] showed a high level of social acceptability for an MAR project in the study area.

The Analytical Framework for Financial Analysis

The study undertakes an analysis to estimate irrigation-related costs and benefits for a typical irrigation farm in the lower Namoi. The analysis considers a cotton irrigation farm, which has three different scenarios for the storage of flood water: surface storage in farm dams, aquifer storage using basin infiltration and ASR using existing wells. All of the surface water allocations, including flood water, is stored in farm dams before application to the fields. Owing to limited water availability, less than 20% of the available land is irrigated, and irrigated land in each year is variable. Irrigated cotton Bt (*Bacillus thuringiensis*) and faba bean *(Vicia faba L)* are the sustainable summer and winter rotations that provide the highest net income per megalitre (ML) of irrigation water applied [39]. It is assumed in the analysis that all required irrigation infrastructure, such as surface storage and the irrigation water delivery network, are already built for the entire irrigation land, as this is a common practice in the study area. The annual irrigation water allocation from all sources for an average cotton farm in the lower Namoi is approximately 1350 ML. However, in this analysis, we only consider and report irrigation costs and returns of 200 ML of flood water, which is only 25% of recent statutory flood water allocations in the study area. The analysis assumes 40% evaporative losses, taking into account current estimates in the study area [37].

Storage and recovery of water underground will require new infrastructure and additional costs, as reported in Section 4.3. Farm economic data, such as the variable cost of farm inputs, cotton prices and gross margins from irrigated and dryland, are adopted from Powell and Scott [39]. The analysis only considers farm-related costs and revenue and does not monetize any socio-economic or environmental cost or revenue that may occur as a result of a change in the water storage option.

Infiltration and Injection Rates That Can be Possible in Lower Namoi

Infiltration and injection rates can highly affect the usefulness of any aquifer recharge and storage facility. Bouwer [40] provides typical infiltration rates for surface infiltration systems in the range from 0.3 to 3 m per day (m/day) with relatively clean and low turbidity river water. For systems that are operated year-round, long-term infiltration rates vary from 30 m/year to 500 m/year, depending on soil type, water quality and climate. In the lower Namoi, the infiltration rate of 0.2 m/day is considered to be likely achieved in many locations.

ASR can achieve injection rates from 0.5 to 8 megalitres per day (ML/day) per borehole (1 megalitre = 1000 cubic meter = 0.8107 acre foot). In the absence of accurate well injection rates based on field monitoring, Pyne [41] observed that injection rates increased with increasing aquifer transmissivities. For the lower Namoi, Williams *et al.* [42] reported that the alluvial aquifers that are primarily tapped for irrigation extraction are associated with the semi-confined Gunnedah and Cubbaroo formations and have transmissivities in the range of 1000–2000 square meters per day (m^2day^{-1}). The yields from bores tapping these aquifers vary up to 250 litres per second in the Gunnedah Formation at depths of 60–90 m and in the deep Cubbaroo Formation at depths of 80–120 m. The shallow Narrabri Formation has transmissivities less than 250 m^2 day^{-1}. For this study, an assumed injection rate of 25 L per second (2.2 ML/day) is considered likely for an ASR well.

Estimation of Costs and Benefits

Cost estimates of aquifer recharge are scarce and can vary considerably with location. Itemized costs for this study were estimated by combining current market rates of earthworks, services and materials for water infrastructure projects in Australia and were adjusted to the local situation in the lower Namoi. Cost estimates were also compared with published data and technical reports of Khan *et al.* [12], Dillon *et al.* [10] and Pyne [13].

Capital costs of basin infiltration were estimated by assuming an infiltration rate of 0.2 m/day and calculating the required land area to achieve 2 ML of recharge per day. The target volume of harvested flood water of 200 ML would, on average, appear in four or more events in a flood year. An infiltration pond with a surface area of 1 ha and an infiltration rate of 0.2 m/day would recharge 50 ML of floodwater in a single cycle of 25 days. The size of the basin here has therefore been designed to operate only for 100 days, in 4 cycles of 25 days each, allowing rest and maintenance. The analysis assumed 40% evaporative losses from surface storage and a 5% MAR loss rate. The MAR loss rate is the percent of water lost during aquifer recharge and recovery from basin infiltration and ASR and can be expressed as:

$$\text{MAR loss rate} = \left(1 - \frac{\text{groundwater volume re cov ered}}{\text{initial water volume used for storage}}\right)\%$$

In the base case, surface storage of flood water, the costs considered are the cost of harvesting 200 ML of flood water and the cost of farm dam annual maintenance. The capital cost of basin infiltration includes the cost of earth works and pipes. Ongoing costs include operation and maintenance of water harvesting and recovery and the cost of basin annual maintenance. An existing bore is assumed to be available for recovery after basin infiltration or for injection and recovery in ASR. The capital cost of an ASR facility on existing farms with a bore primarily includes installing a coagulation and filtration pre-treatment facility. Ongoing operation and maintenance costs for ASR include well maintenance, flood water harvesting, water treatment and water recovery. The analysis assumed a 30-year lifespan for surface storage and basin infiltration and 20 years for ASR, with a 7% uniform discount rate for all options. All capital cost estimates are exclusive of land value. Table 1 summarizes the levelised costs of 200 ML of flood water with each water storage option. Levelised costs are annual unit costs obtained by amortising capital cost components over their expected working life, adding the annual operation, maintenance and management cost and dividing by the annual volume of supply, as defined in Dillon et al. [10].

Table 1. Levelised costs ($/ML) of surface storage and MAR methods in lower Namoi. Adapted from Dillon and Arshad [43]. ASR, aquifer storage and recovery; ML, megalitre

Cost component	Surface storage	Basin infiltration	ASR using existing well
Annual cost of capital items ($/ML)	0.0	32.2	26.0
Annual cost of operation, maintenance and management ($/ML)	22.5	90.5	221.8
Total annual cost ($/ML)	22.5	122.7	247.7

With the additional water saved through MAR, farmers in our example have the choice to irrigate additional land with cotton, faba bean or some combination of the two crops that yields the highest returns. Value brought by the MAR water under each option is estimated from the useable volume of flood water, after evaporative and recovery losses, times the gross margin per ML of mixed cropping of cotton and faba bean on equal land areas. On average, for a typical lower Namoi irrigation farm, the average gross margins for cotton and faba bean are estimated as 310 $/ML and 435 $/ML, respectively. It is assumed that cotton and faba bean are planted on the same land area, as they are summer and winter crops, respectively. Allocating the water accordingly yields an average gross margin of 342.3 $/ML and a net margin of 230 $/ML after subtracting overhead costs. In the analysis, we assume that additional irrigation with the saved water is not going to increase the overhead cost, as the farm size is large enough (1200 ha) and irrigated land cropped each year is variable depending on water availability. In this analysis, we use gross margins as the irrigation returns, which is the total revenue minus the variable cost of production. Table 2 presents the value of crop that can be grown with the useable volume in each case.

Table 2. Irrigation benefits: value of the crop under each water storage option. Adapted from Powell and Scott [39] and Arshad *et al.* [44]

Project component	Surface storage	Basin infiltration	ASR using existing well
Initial volume taken from flooding river ML	200	200	200
Useable volume (after losses) (ML)	120	190	190
Gross value of crop ($/ML)	342.3	342.3	342.3
Irrigation benefits: value of the crop that can be grown with the useable volume in each case ($) (available water times gross margins $/ML)	41,070.6	65,028.4	65,028.4

Results of Financial Cost-Benefit Analysis

A long-term trajectory of the difference of the discounted benefits and discounted costs of the three water storage options is expressed in Figure 3 as net present value using the fixed data in Table 2.

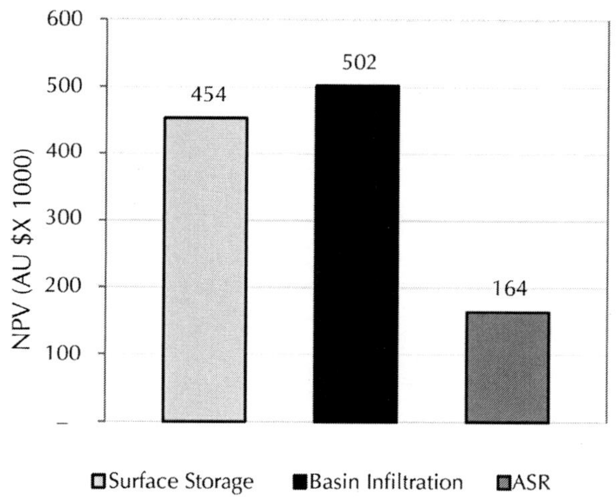

Figure 3. Net present value (NPV) of surface storage, basin infiltration and aquifer storage and recovery options.

The results show that MAR using the basin infiltration method will yield 11% more value than surface storage of irrigation water. ASR using existing wells appears to be uneconomical, with 64% less value than surface storage, mainly due to the high capital and water treatment costs required for an ASR system.

The cost and additional value of basin infiltration is highly dependent on the infiltration rates; as infiltration rates increase, the capital costs decrease, and the value of saved water increases. Conversely, as infiltration rates decrease, the capital cost increases, and the additional value of basin infiltration decreases. With a reduction in the infiltration rates, a cross-over point is reached, where the additional value brought by basin infiltration becomes zero and its NPV is exactly equal to that of surface storage. The following section expands the analysis to explore cross-over points of infiltration rates and other variables.

IDENTIFICATION OF CROSS-OVER POINTS IN A SINGLE VARIABLE

In single variable analysis, the aim is to identify how far a single variable needs to change to reach a cross-over point for the two compared options. A cross-over point may not always exist; there might be situations where the cross-over point falls outside the minimum or maximum limits considered for the analysis or when the change in the cost or benefit is in the same direction. Such a situation is noted with the use of the acronym, NA, for not applicable, in the tables and following text. A cross-over point for basin infiltration and surface storage occurs when their NPVs are equal; and similarly, for ASR and surface storage, as well as basin infiltration and ASR. Figure 1 showed the cross-over point for pumping cost. Figure 4 illustrates cross-over points for basin capital cost.

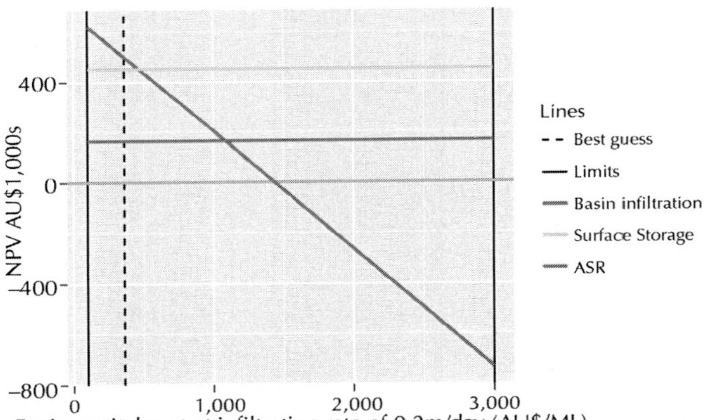

Figure 4. NPV for varying basin capital cost in three water storage scenarios, showing cross-over points at intersections between lines.

A cross-over point between basin infiltration and surface storage is possible when the basin capital cost increases from 363 \$/ML to 466.69 \$/ML. That increase in the capital cost will equate to the NPV of the two compared options. Similarly, a cross-over point between basin infiltration and ASR is possible when the basin capital cost increases from 363 \$/ML to 1085.55 \$/ML. No cross-over point is identified between surface storage and ASR (it is NA). The increase in the basin capital cost may result from increases in the price of services and materials or the need to construct a larger pond due to a reduction in infiltration rates. Rather than drawing these curves for every variable, the values of the cross-over points are reported in Table 3 and discussed in the following text.

Table 3 lists cross-over points for 14 variables when each is varied separately. These cross-over points represent the minimum requirements for MAR to be preferred to surface water storage, assuming that the values of other variables listed in the table remain fixed. For example basin infiltration is financially better than surface storage when pumping cost does not exceed 53.63 $/ML or the surface evaporation rate does not fall below 34%, and so on. The variables selected are the most important when undertaking a financial comparison of surface storage with the two MAR options. In the following section, we discuss the basis of how these cross-over points may be reached in reality for each single variable.

Table 3. Single variable cross-over points in three scenarios

No.	Variable (Unit)	Cross-Over Point			
		Best Guess (Modelled) Value	**Surface Storage and Basin Infiltration**	**Surface Storage and ASR**	**Basin Infiltration and ASR**
1	Pumping cost ($/ML)	35	53.63	NA	NA
2	Surface evaporation rate (%)	40	34	74	NA
3	Basin capital cost ($/ML)	363	466.69	NA	1,085.55
4	Basin infiltration rate (m/day)	0.2	0.16	NA	0.07
5	Basin maintenance rate (% of capital cost)	10	15	NA	NA
6	MAR loss rate (% of target storage volume)	5	11	NA	NA
7	ASR water treatment cost ($/ML)	150	NA	13.25	NA

8	ASR maintenance rate (% of capital cost)	0.07	NA	NA	NA
9	Price of cotton ($/ bale)	538	475.64	1,155.22	NA
10	Price of faba bean ($/tonne)	348	229.52	NA	NA
11	Discount rate (%)	7	13	NA	NA
12	Lifespan of surface storage (Year)	30	48.16	5.57	NA
13	Lifespan of basin infiltration (Year)	30	23.51	NA	6.69
14	Lifespan ASR (Year)	20	NA	NA	NA

Discussion of Single Variables

Pumping Costs and Surface Evaporation Rates

A cross-over point between surface storage and basin infiltration is possible when pumping costs increase by 53% to become 53.63 $/ML; an increase in the cost of pumping will cause an increase in the cost of agricultural production and a decrease in farm benefits (NPV) from basin infiltration. A cross-over point between basin infiltration and ASR is NA, because the rate of increase in pumping cost applies to both aquifer storage options. Similarly, there is no cross-over point between surface storage and ASR, as the lowest possible pumping cost considered in the analysis (6.25 $/ML) will not make ASR financially superior or equal to surface storage.

Low surface evaporation rates will make surface storage financially superior to MAR, as less water will be lost from surface storage, making more water available and resulting in larger benefits. A cross-over point between surface storage and basin infiltration is possible when evaporation rates decrease by 15%, from 40%, to become 34%. For evaporation rates, a cross-over point between surface storage and ASR is possible when evaporation rates increase to 74%, whereas the cross-over point between basin infiltration and ASR is NA.

Basin Capital Cost, Basin Infiltration Rate and Basin Maintenance Rate

An increase in basin capital cost will increase the overall cost and lower the benefits with a concomitant decrease in NPV. For the basin capital cost, a cross-over point between surface storage and basin infiltration is possible when the capital cost of basin infiltration increases from 363 $/ML to 466.69 $/ML.

A decrease in the infiltration rates will recharge less water per unit area of infiltration basin, requiring a large infiltrating pond area with larger capital cost, or with decreased infiltration rates, less water will infiltrate and be stored underground. A cross-over point between surface storage and basin infiltration is possible when infiltration rates drop from 0.2 m/day to 0.16 m/day. Similarly, a cross-over point between basin infiltration and ASR is achieved when infiltration rates drop from 0.2 m/day to 0.07 m/day. An increase in the basin maintenance rates will increase the overall cost of basin infiltration, reducing NPV in comparison to the compared options. A cross-over point between surface storage and basin infiltration is possible when basin maintenance rates increase from 10% to become 15%. The three considered variables do not apply when comparing surface storage and ASR, such that the corresponding cross-over points are NA.

MAR Loss Rate

Increasing the MAR loss rate makes MAR financially less attractive, because it reduces the volume of water recovered and the resulting benefits, though some pumping cost is saved, as less water is recovered with an increase in the MAR loss rate. In other words, a higher MAR loss rate represents a lower recoverability and, therefore, lower useful storage [22, 45]. For benefits to be realized, the volume of water that is not recovered from storage must be less than evaporation losses. This applies to both MAR methods when compared to surface storage. A cross-over point between basin infiltration and surface storage is possible when the MAR loss rate reaches 11%. A cross-over point between basin infiltration and ASR is NA.

ASR Water Treatment Cost and ASR Maintenance Rates

A cross-over point for ASR maintenance rate is not possible when ASR is compared with surface storage and basin infiltration. Even its cheapest possible value, when considered alone, does not achieve an NPV equal or superior to basin infiltration and surface storage. The ASR water treatment cost only has a cross-over point if the treatment cost decreases by 91% to 13.25 $/ML. Increases in both variables increase the cost of ASR and, hence, (further) diminish its advantage over the other options.

Price of Cotton and Faba Bean

A decrease in the price of cotton and faba bean will influence the benefits of all three water storage options and lower NPVs for each case. A cross-over point for the price of cotton and the price of faba bean between surface storage and basin infiltration is possible when the price of cotton drops from $538 per bale to $475.64 per bale, and the price of faba bean drops from $348 per tonne to $229.52 per tonne, which are 11% and 34% drops from the best guess values, respectively. A cross-over point for the cotton and faba bean price is possible between surface storage and ASR when the price of cotton rises to $1,155.22 per bale, an increase of 114%. No cross-over point between surface storage and ASR is possible with the highest price considered possible for faba beans.

Discount Rate and Project Lifespan

An increase in the discount rate tends to increase the levelised cost of the two MAR options, in particular through the basin capital cost and the capital cost of establishing an ASR treatment facility. This will result in lower NPVs from the two MAR options. A cross-over point between surface storage and basin infiltration is possible at a discount rate of 13%, while there is no cross-over point between surface storage and ASR. Because ASR is already more expensive than surface storage, a higher discount rate will make ASR even more expensive, while the lowest considered discount rate of 1% will not be able to raise the NPV of ASR to be equal or superior to surface storage. Similarly, a lower discount rate will make basin infiltration more favourable than ASR, so no cross-over point is possible.

Lowering the lifespan of an option increases its levelised cost, such that the NPV of that particular option is lowered. A cross-over point between surface storage and basin infiltration is possible when the lifespan of surface storage increase from 30 years to 48.16 years or the lifespan of the basin infiltration drops from 30 years to become 23.51 years. Similarly, a cross-over point between surface storage and ASR exists when the lifespan of surface storage drops to 5.57 years. A cross-over point between basin infiltration and ASR is possible when the lifespan of the basin infiltration drops to 6.69 years. No cross-over point for the lifespan of ASR is possible when compared with basin infiltration and with surface storage options.

Changes in Cross-Over Points Due to Interactions between Variables

The values at which cross-over points occur are affected by the values of other variables, so it is important to consider interactions between variables. Every variable that either increases or decreases changes the financial advantage of MAR in comparison to surface storage. We describe the advantage of MAR in terms of change in the position (value) of cross-over points with respect to the best guess. The interaction of two variables can bring a cross-over point closer or further to the best guess. Two variables can interact in a way that they can increase, decrease or balance the effect of each other on the resulting advantage of MAR, depending on whether changes in the variable increase or decrease the financial advantage of MAR.

A cross-over point that moves away from the best guess value indicates increasing financial advantage for MAR. Conversely, when it moves closer to the best guess, the financial advantage decreases. The movement of a cross-over point closer to the best guess reveals situations where the benefits of MAR are reduced and could ultimately have equal benefits to surface storage when the cross-over point coincides with the best guess value.

Figure 5 and Figure 6 illustrate with examples where the advantage of MAR over surface storage changes due to the interaction of variables.

This is expressed through changes in the cross-over point of the MAR loss rate.

Given that increased costs reduce the relative benefit of MAR, when costs increase, the cross-over point for the MAR loss rate moves closer to the best guess value (Figure 5). Similarly, lower prices of crops decrease the benefit of MAR, and the cross-over point moves closer to the best guess (middle bar in Figure 6). When costs and prices both increase, the cross-over point can move closer or further from the best guess, depending on the level of change in costs and prices (bottom bar in Figure 6).

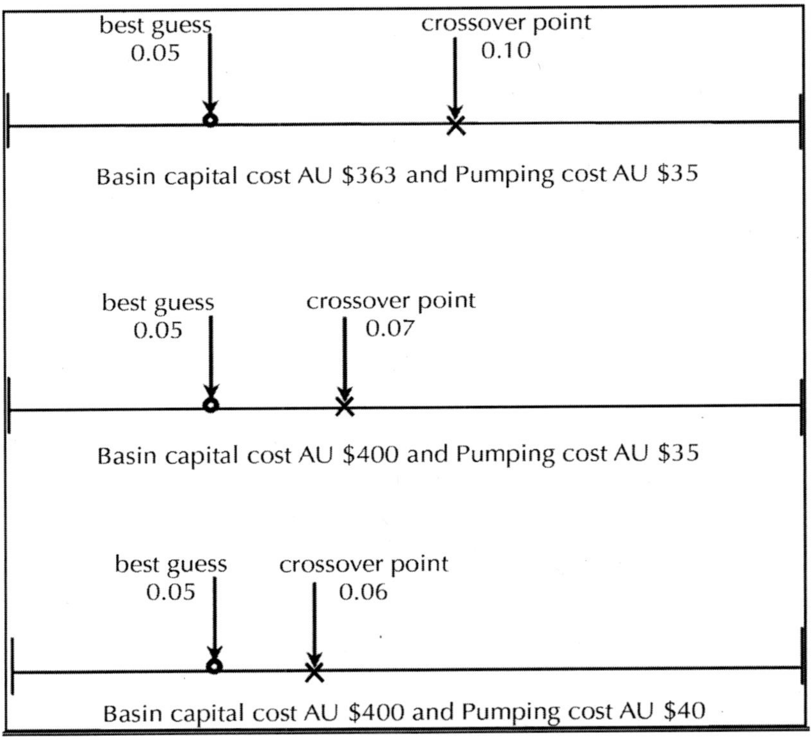

Figure 5. Plot of the MAR loss rate when costs increase. An example of a cross-over point moving toward the best guess.

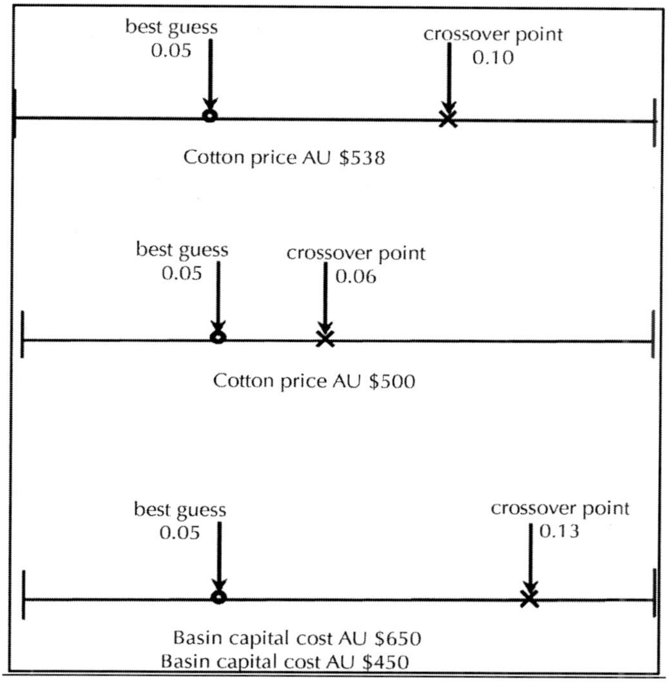

Figure 6. Plot of the MAR lose rate when costs and prices change. An example of the cross-over point changing position when costs and prices both increase.

Assessing the Risk of Attaining Cross-Over Points

Uncertainty in the financial assessment of MAR can be assessed by evaluating whether the scenarios described by the cross-over points identified are likely to be experienced in reality. If this occurs, then MAR may not be financially attractive. Alternatively, other measures may need to be taken to avoid situations leading to the cross-over point. Note that initial financial analysis suggests that basin infiltration is a favourable investment. As mentioned in the Introduction, the aim of this analysis is therefore to play the devil's advocate, that is to systematically search for reasons that requirements may not be met and that failure might occur.

While cross-over points could be assessed probabilistically, a simple approach is to say that a cross-over point is of greater concern if it is closer to the best guess value. This implies that investment in the MAR infrastructure is at greater risk of not making additional profits than surface storage because the return from MAR becomes closer to that of surface storage. On the other hand, the value of a cross-over point may fall outside the bounds (minimum and maximum limits) that are considered to be of concern, in which case, the analysis suggests that the minimum requirements will be met.

Following this approach, Table 4 shows the cross-over point of greatest concern when surface storage and basin infiltration are compared. The point was identified by simultaneously varying all of the variables and searching for a combination where each variable is closest to the best guess, relative to bounds. The bounds were defined by the authors based on an understanding of the factors influencing the variables, taking into account the expected variability, considering the lack of complete the knowledge of hydrogeological variables and the actions that can be taken to manage these concerns. In interpreting the results, the combination of values is assessed, not just each variable separately, and the reasons for the bounds selected are explained.

Table 4. Cross-over point of greatest concern with basin infiltration vs. surface storage, using a subset of variables

	Variable	**Minimum Bound**	**Maximum Bound**	**Best Guess**	**Point of Greatest Concern**	**Change from Best Guess**
1	Pumping cost ($/ML)	6.25	225	35	37.22	2.22
2	Surface evaporation rate (%)	10	100	40	40	0
3	Basin capital cost ($/ML)	100	3,000	363	393.82	30.82
4	Basin infiltration rate (m/day)	0.01	2	0.2	0.2	0
5	Basin maintenance rate (% of capital cost)	1.0	40	10	10	0
6	MAR loss rate (% of target storage volume)	0	85	5	6	1

7	Price of cotton ($/bale)	50	1500	538	532.30	−5.70
8	Price of faba bean ($/tonne)	50	1400	348	344.52	−3.48
9	Discount rate (%)	1	50	7	8	−1
10	Lifespan of surface storage (Year)	2	50	30	30.23	0.23
11	Lifespan of basin infiltration (Year)	2	50	30	29.67	−0.33

Table 4 shows that the values of cross-over points are very close to the best guess and, hence, are of concern. The point of greatest concern describes a scenario of particularly unfavourable conditions, namely when all of the variables interact and change simultaneously. The scenario of greatest concern describes a situation where pumping costs have increased and the prices of cotton and faba bean have decreased. Basin capital cost turns out to be higher than expected, as well as the MAR loss rate. The lifespan of the basin infiltration project is marginally shorter than that of a surface storage project. Other variables remain at their best guess.

Individually, all variables of the scenario appear to be of great concern. However, in reality it is unlikely that all variables change at once and result in the situation described in Table 4. We analyse groups of variables to assess whether or not the generated scenario is possible, what mitigation options might prevent this cross-over point from occurring and what adaptation actions might be taken if it the scenario described by the cross-over point does occur.

Pumping Costs and Surface Evaporative Rates

The cross-over point of this variable is very close to the best guess value and, hence, may be reached. Based on historical trends, energy costs are expected to increase in the future, despite efficiency improvements in pumping technologies. However, the effect of higher pumping costs may be balanced or outweighed if there is an increase in the price of cotton and faba bean in the future. In addition, if farming becomes uneconomical at some stage, it is possible that government might provide subsidies for pumping to maintain agricultural production and preserve the livelihoods of farmers. Using alternate sources of energy, such as wind and solar, can be cheaper mitigation options in the future.

High head gravity feed systems can be designed in certain cases to avoid pumping costs [9].

Surface evaporative rates are expected to increase under climate variability and change [46]. Evaporative rates may also be higher for farms where surface storage is shallow and depending on the water colour and turbidity. Higher evaporative rates will favour MAR, so this is unlikely to be a reason not to proceed with MAR. Reducing evaporative losses from surface storage at costs cheaper than those of setting up a basin infiltration system could have been a reason not to proceed with basin infiltration.

Basin Capital Cost, Infiltration Rate and Basin Maintenance Rate

The increase in the basin capital cost seems likely to occur if the investment is delayed, as the cost of labour, construction materials and energy prices for earth moving machinery are expected to rise due to inflation and other economic factors. Similarly, the values of basin infiltration rates and basin maintenance rates exactly coincide with the current best guess estimates, and hence, the two variables are of great concern. The infiltration rate is a function of a number of variables, with water quality a major factor.

A few mitigation options exist to avoid increases in basin capital cost. Field trials and geophysical investigations can help find suitable sites where high infiltration rates can be achieved. Basin maintenance is related to the amount of silt and other suspended and organic matter contained in the floodwater. Basins can be sized to allow rest and maintenance. In the lower Namoi catchment, floodwater already passes through a *de facto* two-stage sediment and silt removal process. Firstly, it is retained in large public dams before release, thereby reducing heavy silt loads; secondly, at the farm level, floodwater is kept in farm dams as temporary storage before recharging begins. The two-step sediment removal process can be advantageous in lowering the cost of basin maintenance.

MAR Loss Rate

In the lower Namoi, more than four decades of groundwater pumping have dropped the water levels, and in many places, rivers and streams (naturally) recharge groundwater [47], such that useful storage exists at a large scale. At the farm scale, while water may not physically stay within a farmer's land and, as such, is not physically stored, the system of surface and groundwater water rights means that injected or infiltrated water could, in principle, be allocated to the farm anyway, in a form of "regulatory storage" [22]. This results in potentially extremely high recovery rates (95%) and low loss rates, as a farmer benefits from contributing water to a common pool rather than being restricted to physically retrieving the water that they recharged. The loss rate determined by regulation could however be affected by a number of broad-scale issues. For example, the MAR loss rate can be of concern for locations where surface water and groundwater connectivity exists and where streams and rivers gain groundwater from aquifers, which is rare in lower Namoi. Low recovery is possible only in aquifers that contain brackish or high salinity water, due to the mixing of fresh recharge water with the native high salinity groundwater. This may occur in some parts of the lower Namoi, particularly areas where drops in groundwater hydraulic heads have resulted in the mixing of saline and freshwater within different layers of aquifers. In areas of excessive groundwater extraction, groundwater hydraulic heads can drop and allow saline water to enter into pumping wells [42], thereby increasing the salinity levels of the recovered water and resulting in less recovery of the volume of freshwater recharged initially.

Price of Cotton and Faba Bean

Cross-over points for the price of cotton and faba bean are not likely to occur, and they are not of greater concern. The future price of cotton is expected to remain stable or increase because of ongoing demand and an established linkage of the Australian cotton industry to overseas markets, where demand exists and can be expected to grow. In the future, with limited irrigation water availability at the global scale, international prices of cotton are expected to rise, rather than decrease. Other cotton producing and competing countries, such as China, Pakistan and Egypt, are likely to become more water stressed in future.

Additionally, with world population growth continuing unabated, a higher demand for cotton is expected. The price of faba bean is also expected to increase in the future; however, a drop in the price of faba bean is also possible whenever supply exceeds the local demand. A change in the price of faba bean is not a major concern, because it is a local crop mainly used for cattle and human consumption and has limited potential for export in national and international markets. Faba bean is not a major source of farm revenue, and if at some point, there is an oversupply and a drop in price occurs, faba bean can be replaced with some other high value crop. Any rise in the sale price of both cotton and faba bean would also compensate for increases in pumping costs and other MAR infrastructure costs.

Discount Rate and Lifespan of Projects

A 7% discount rate coincides with the current best guess and is highly likely to occur and is therefore of great concern. Discount rates of more than 7% will make MAR financially un-attractive. As this may occur if the cost of borrowing capital is high, farmers may search for financing at lower rates or governments may assist farmers to set up special MAR grants or loans involving the least possible interest rates. Cross-over points for the lifespan of surface storage and basin infiltration almost coincide with the best guess (30 years) and are of great concern. The lifespan of basin infiltration can be enhanced by drying of basins, frequent scarping of accumulated silt layers and controlling weed growth.

CONCLUSIONS

Break-even analysis of cross-over points is one way of assessing the financial performance of MAR under uncertainty. Cost-benefit analysis of surface storage and MAR helps to compare options in financial terms, but results cannot be relied upon completely without due consideration of uncertainty. Our approach to addressing uncertainty is to undertake a financial cost-benefit analysis by analysing a range of values for influencing variables and to establish thresholds (cross-over points) where financial returns from surface storage and MAR are equal. Once the thresholds are established, mitigation options can be identified and put in place to avoid variables reaching identified thresholds.

The analysis of cross-over points can be undertaken to identify minimum requirements under which MAR can be more advantageous than surface storage, and this was illustrated for the lower Namoi. For this catchment, MAR using basin infiltration can be financially superior to surface storage, but this depends on the selection of a suitable site where a high infiltration rate, low loss rates and other minimum requirements can be achieved. Further exploration of MAR through field trials and geo-physical investigation is suggested in areas of lower Namoi. MAR can be a potential option to achieve future water supply goals, provided that it is technically feasible and more financially viable than surface storage.

ACKNOWLEDGMENTS

This research was partly funded by the Australian Government's Endeavour Scholarships and Fellowships programme, the National Centre for Groundwater Research and Training (NCGRT), Australia. The authors thank Tony Jakeman and Michael Asher of Australian National University and two reviewers for their comments on this paper. We also thank Ejaz Qureshi of the Australian Centre of International Agriculture Research (ACIAR), Peter Dillon from the Commonwealth Scientific and Industrial Research Organisation (CSIRO) and David Pyne, an independent consultant and pioneer of ASR technology, for discussions of the topic in personal communications.

REFERENCES

1. Wisser, D.; Frolking, S.; Hagen, S.; Bierkens, M.F. Beyond peak reservoir storage? A global estimate of declining water storage capacity in large reservoirs. *Water Resour. Res.* 2013, *49*, 5732–5739.
2. Konikow, L.F.; Kendy, E. Groundwater depletion: A global problem. *Hydrogeol. J.* 2005, *13*, 317–320.
3. Giordano, M. Global groundwater? Issues and solutions. *Annu. Rev. Environ. Resour.* 2009, *34*, 153–178.

4. UN Water. The United Nations World Water Development Report 2014: Water and Energy. UNESCO: Paris, France, 2014.
5. Shiklomanov, I.A. Appraisal and assessment of world water resources. *Water Int.* 2000, *25*, 11–32.
6. Gale, I.; Neumann, I.; Calow, R.; Moench, M. The effectiveness of Artificial Recharge of groundwater: A review. In *British Geological Survey, Commercial Report CR/02/108N*; British Geological Survey: Nottingham, UK, 2002.
7. Gale, I.; Macdonald, D.; Calow, R.; Neumann, I.; Moench, M.; Kulkarni, H.; Mudrakartha, S.; Palanisami, K. *Managed Aquifer Recharge: An Assessment of Its Role and Effectiveness in Watershed Management: Final Report for DFID KAR Project R8169, Augmenting Groundwater Resources by Artificial Recharge: AGRAR*; British Geological Survey: Nottingham, UK, 2006.
8. Dillon, P. Future management of aquifer recharge. *Hydrogeol. J.* 2005, *13*, 313–316.
9. Missimer, T.M.; Guo, W.; Maliva, R.G.; Rosas, J.; Jadoon, K.Z. Enhancement of wadi recharge using dams coupled with aquifer storage and recovery wells. *Environ. Earth Sci.* 2014, *June*.
10. Dillon, P.; Pavelic, P.; Page, D.; Beringen, H.; Ward, J. Managed aquifer recharge. In *An introduction Waterlines Report Series No. 13*; National Water Commission: Canberra, Australia, 2009.
11. Donovan, D.J.; Katzer, T.; Brothers, K.; Cole, E.; Johnson, M. Cost-benefit analysis of artificial recharge in Las Vegas Valley, Nevada. *J. Water Resour. Plan. Manag.* 2002, *128*, 356–365.
12. Khan, S.; Mushtaq, S.; Hanjra, M.A.; Schaeffer, J. Estimating potential costs and gains from an aquifer storage and recovery program in Australia. *Agric. Water Manag.* 2008, *95*, 477–488.
13. Pyne, R.D.G. Aquifer Storage Recovery: Economics and Recent Technical Advances in Achieving Ground Water Supply Sustainability & Reliability through Managed Aquifer Recharge. In Proceedings of the 7th International sumposium on Managed Aquifer Recharge, Abu Dahbi, IAH, United Arab Emirates, 9–13 October 2010.
14. Griffin, R.C. *Water Resource Economics*; MIT Press: Cambridge, MA, USA, 2006.

15. Kalantari, N.; Rangzan, K.; Thigale, S.S.; Rahimi, M.H. Site selection and cost-benefit analysis for artificial recharge in the Baghmalek plain, Khuzestan Province, southwest Iran. *Hydrogeol. J.* 2010, *18*, 761–773.
16. Maliva, R. Economics of managed aquifer recharge. *Water* 2014, *6*, 1257–1279.
17. Almansa, C.; Martínez-Paz, J.M. What weight should be assigned to future environmental impacts? A probabilistic cost benefit analysis using recent advances on discounting. *Sci. Total Environ.* 2011, *409*, 1305–1314.
18. Schwenk, C.R.; Cosier, R.A. Effects of the expert, devil's advocate, and dialectical inquiry methods on prediction performance. *Organ. Behav. Hum. Perform.* 1980, *26*, 409–424.
19. Gutteridge Haskins & Davey. Australian Groundwater Technologies. In *Feasibility of Managed Aquifer Recharge for Agriculture*; Waterlines Report 45. National Water Commission: Canberra, Australia, 2011.
20. Rawluk, A.; Curtis, A.; Sharp, E.; Kelly, B.F.J.; Jakeman, A.J.; Ross, A.; Arshad, M.; Brodie, R.; Pollino, C.A.; Sinclair, D.; *et al.* Managed aquifer recharge in farming landscapes using large floods: An opportunity to improve outcomes for the Murray-Darling Basin? *Australas. J. Environ. Manag.* 2012, *20*, 1–15.
21. Bouwer, H. Artificial recharge of groundwater: Hydrogeology and engineering. *Hydrogeol. J.* 2002, *10*, 121–142.
22. Maliva, R.G.; Missimer, T.M. *Aquifer Storage and Recovery and Managed Aquifer Recharge Using Wells: Planning, Hydrogeology, Design and Operation*; Schlumberger: New York, NY, USA, 2010.
23. Pedretti, D.; Barahona-Palomo, M.; Bolster, D.; Fernández-Garciaz, D.; Sanchez-Vila, X.; Tartakovsky, D.M. Probabilistic analysis of maintenance and operation of artificial recharge ponds. *Adv. Water Resour.* 2011, *36*, 23–35.
24. Holländer, H.; Mull, R.; Panda, S. A concept for managed aquifer recharge using ASR-wells for sustainable use of groundwater resources in an alluvial coastal aquifer in Eastern India. *Phys. Chem. Earth Parts A/B/C* 2009, *34*, 270–278.

25. Pyne, R.D.G. *Groundwater Recharge and Wells: A Guide to Aquifer Storage Recovery*; CRC Press: Boca Raton, FL, USA, 1995.
26. Ross, A.; Arshad, M. Underground water storage and banking: Benefits, barriers and ways forward. In Proceedings of the 8th International Symposium on Managed Aquifer Recharge, ISMAR 8, Beijing, China, 15–19 October 2013.
27. Mohamed, S.; McCowan, A.K. Modelling project investment decisions under uncertainty using possibility theory. *Int. J. Proj. Manag.* 2001, *19*, 231–241.
28. Jovanović, P. Application of sensitivity analysis in investment project evaluation under uncertainty and risk. *Int. J. Proj. Manag.* 1999, *17*, 217–222.
29. Christopher Frey, H.; Patil, S.R. Identification and Review of Sensitivity Analysis Methods. *Risk Anal.* 2002, *22*, 553–578.
30. Rockström, J. Water for food and nature in drought–prone tropics: Vapour shift in rain–fed agriculture. *Philos. Trans. R. Soc.Lond. Ser. B Biol. Sci.* 2003, *358*, 1997–2009.
31. Pinheiro, J.; Bates, D.; DebRoy, S.; Sarkar, D.; R Development Core Team. *Nlme: Linear and Nonlinear Mixed Effects Models*; R Package Version 3.1–97. R Foundation for Statistical Computing: Vienna, Austria, 2011.
32. Brent, R.P. *Algorithms for Minimization without Derivatives*; Courier Dover Publications: New York, NY, USA, 1973.
33. Guillaume, J.H.A.; Arshad, M. Mar-under-uncertainty: Code for MAR analysis in Water journal article. Available online: http://dx.doi.org/10.5281/zenodo.11704 (accessed on 12 September 2014).
34. Murray-Darling Basin Authority. *The proposed Groundwater Baseline and Sustainable Diversion Limits: Methods Report*; MDBA publication No: 16/12. Murray-Darling Basin Authority: Canberra, Australia, 2012.
35. Smithson, A. *Lower Namoi Groundwater Source: Groundwater Management Area 001 Groundwater Status Report 2008*; NSW Department of Water and Energy: Sydney, Australia, 2009.
36. Australian Cotton Growers. A bi-monthly magazine providing technical and marketing information to the Australian cotton

industry, 2012. Available online: http://www.greenmountpress.com.au/cottongrower (accessed on 15 September 2014).

37. Wigginton, D. Measuring losses from on-farm water storages in the cotton industry. In Proceedings of the Irrigation 2011-New Horizons, Fresh Ideas Regional Conference and Exhibition, Irrigation Australia Ltd, Sydney, Australia, 22–25 August 2011.

38. Arshad, M.; Jakeman, A.J.; Kelly, B.F.J.; Qureshi, M.E.; Croke, B.F.W.; Blakers, R.S.; Ross, A.J.; Curtis, A.; Rawluk, A. Assessing the Potential of Managed Aquifer Recharge: Preliminary Hydrological Findings from a Scoping Study in the Lower Namoi Valley, Australia. IGWC-2012. In Proceedings of the Fifth International Groundwater Conference on the Assessment and Management of Groundwater Resources in Hard Rock Systems with Special Reference to Basaltic Terrain, Department of Geology, Maulana Azad College, Aurangabad, India, 18–21 December 2012.

39. Powell, J.; Scott, F. A representative irrigated farming system in the Lower Namoi Valley of NSW: An economic analysis. In *Economic Research Report*; Industry & Investment NSW: Narrabri, Australia, 2011.

40. Bouwer, H. Artificial recharge of groundwater: Systems, design, and management. In *Hydraulic Design Handbook*; McGraw-Hill: New York, NY, USA, 1999; pp. 24.1–24.44.

41. Pyne, R.D.G. Aquifer storage recovery: A guide to groundwater recharge through wells. ASR systems: Liverpool, NY, USA, 2005.

42. Williams, R.M.; Merrick, N.P.; Ross, J.B. Natural and Induced Recharge in the Lower Namoi Valley, New South Wales. In Proceedings of the Symposium on Groundwater Recharge, Mandurah, Australia, 6–9 July 1987.

43. Dillon, P.; Arshad, M. Managed Aquifer Recharge in Integrated Water Resource Management: Chapter 19. In *Integrated Groundwater Management; Concepts, Approaches and Challenges*; Jakeman, A., Barreteau, O., Hunt, R., Rinaudo, J.D., Ross, A., Eds.; Springer: Berlin, Germany Submitted for Publication. .

44. Arshad, M.; Qureshi, M.; Jakemana, A. Cost-benefit analysis of farm water storage: Surface Storage *versus* Managed Aquifer Storage. In Proceedings of the 20th International Congress on

Modelling and Simulation, Adelaide, Australia, 1–6 December 2013.

45. Maliva, R.G. Groundwater banking: Opportunities and management challenges. *Water Policy* 2014, *16*, 144–156.
46. CSIRO. Water availability in the Namoi. In *A Report to the Australian Government from the CSIRO Murray-Darling Basin Sustainable Yields Project*; CSIRO: Clayton South, Australia, 2007; p. 154.
47. Kelly, B.; Merrick, N.; Dent, B.; Milne-Home, W.; Yates, D. *Groundwater Knowledge and Gaps in the Namoi Catchment Management Area*; Report NCGM 2007/1. National Centre for Groundwater Management, University of Technology Sydney: Sydney, Australia, 2007.

Chapter 9

Policy Preferences about Managed Aquifer Recharge for Securing Sustainable Water Supply to Chennai City, India

Norbert Brunner [1], Markus Starkl [1], Ponnusamy Sakthivel [1], Lakshmanan Elango [2], Subbaiah Amirthalingam [1], Chinniyampalayam E. Pratap [1], Munuswamy Thirunavukkarasu [3] and Sundaram Parimalarenganayaki [2]

[1]Centre for Environmental Management and Decision Support (CEMDS), Vienna A-1180, Austria

[2]Department of Geology, Anna University, Chennai 600025, India

[3]SPT Consultancy Services, Chennai 600011, India

ABSTRACT

The objective of this study is to bring out the policy changes with respect to managed aquifer recharge (focusing on infiltration ponds), which in the view of relevant stakeholders may ease the problem of groundwater depletion in the context of Chennai City; Tamil Nadu; India. Groundwater is needed for the drinking water security of Chennai and overexploitation has resulted in depletion and seawater intrusion. Current policies at the municipal; state and national level all support recharge of groundwater and rainwater harvesting to counter groundwater depletion. However, despite such favorable policies, the legal framework and the administrative praxis do not support systematic approaches towards managed aquifer recharge in the periphery of Chennai. The present study confirms this, considering the mandates of governmental key-actors and a survey of the preferences and motives of stakeholder representatives. There are about 25 stakeholder groups with interests in groundwater issues, but they lack a common vision. For example, conflicting interest of stakeholders may hinder implementation of certain types of managed aquifer recharge methods. To overcome this problem, most stakeholders support the idea to establish an authority in the state for licensing groundwater extraction and overseeing managed aquifer recharge.

INTRODUCTION

In India, as well as in many other countries (e.g., China [1]), overexploitation of groundwater is a serious problem. It has caused declining groundwater levels, deterioration of water quality, and in coastal regions intrusion of seawater. Such a situation may lead to a race for pumping water for irrigation, which accelerates groundwater depletion and ends in a "tragedy of the commons" [2,3]. This becomes evident by higher energy costs for pumping irrigation water: In India, energy for farmers is subsidized (diesel, electricity) or given free and the escalation of subsidies for agriculture burdens government budgets [4].

To overcome groundwater depletion and the associated costs, governments may support managed aquifer recharge (MAR). MAR is the purposeful recharge of water to aquifers for subsequent recovery or environmental benefit, such as rainwater harvesting (RWH), infiltration ponds, or check dams. These are considered in this paper, as they generate water supplies from sources that may otherwise be lost due to runoff [5, 6, 7, 8]. MAR also has the aim of preserving or improving groundwater quality. Related groundwater management actions can include substituting alternative water sources for groundwater (the paper considers desalination) and "non-structural policy measures", by which the paper means demand management to promote water conservation (e.g., by water pricing or state sponsored incentive programs to reduce cropping; [9] is an early example).

The basis of this study is a water supply scenario for Chennai, where overexploitation of groundwater has become a threat to drinking water security [7]: "Chronic water shortages mark the norm in this city." Thereby, for the state of Tamil Nadu the legal framework provides a favorable atmosphere for groundwater management, making e.g., RWH on roofs mandatory since 2001. Also present water polices of India are favorable, acknowledging MAR as an important tool for sustaining water supplies for all kinds of users [10]. However, MAR involves multiple agencies, which may not always cooperate or share information [11]. At the same time, there are many different stakeholder groups and their interests in groundwater recharge, groundwater use, or quality of groundwater may not be compatible with each other; rather multiple conflicts of interests (e.g., urban *vs.* peri-urban and rural) are to be expected [12,13].

Therefore, the paper focuses on the perception of these stakeholder groups on MAR, considering the broader context of groundwater management. It also asks, which policy changes the stakeholders deem necessary to implement a specific MAR approach, namely the construction of many small infiltration ponds.

BACKGROUND INFORMATION: CURRENT WATER SUPPLY AND FUTURE OPTIONS

Chennai City (formerly Madras) is the capital of Tamil Nadu state. With 4.7 million people (2011 census) and an area of 426 km^2 it is the sixth largest city in India. (Official statistics refer to the old boundaries prior to the expansion of city limits in the year 2011.) With a larger metropolitan area of 1189 km^2 and nine million people, it is the fourth largest metropolitan agglomeration in India. The Chennai Metropolitan Water Supply & Sewerage Board (CMWSSB), a statutory body established in 1978, is responsible for water supply and sewerage functions. It operates in the city, only, but is expected to gradually extend its services to the entire metropolitan area.

Over 90% of the water supply of Chennai is covered by water stemming from reservoirs, which are depending on the monsoon rains ([14] and Figure 1). When the reservoirs are empty then the water to the city is mostly met by groundwater to cover the gap in water supply. However, due to exploitation of the groundwater resources (pumping of groundwater for domestic, industrial and agricultural water supply), the contribution of groundwater to the water supply of Chennai has diminished, from a maximum of 25% to around 6% during 2000. At the beginning of the 2013 summer season (March 2013) the share of groundwater was as low as 1% [14]. This indicates over-dependence on all current sources to meet Chennai's water supply. Further, the decline of the groundwater level has led to the intrusion of seawater in the coastal area.

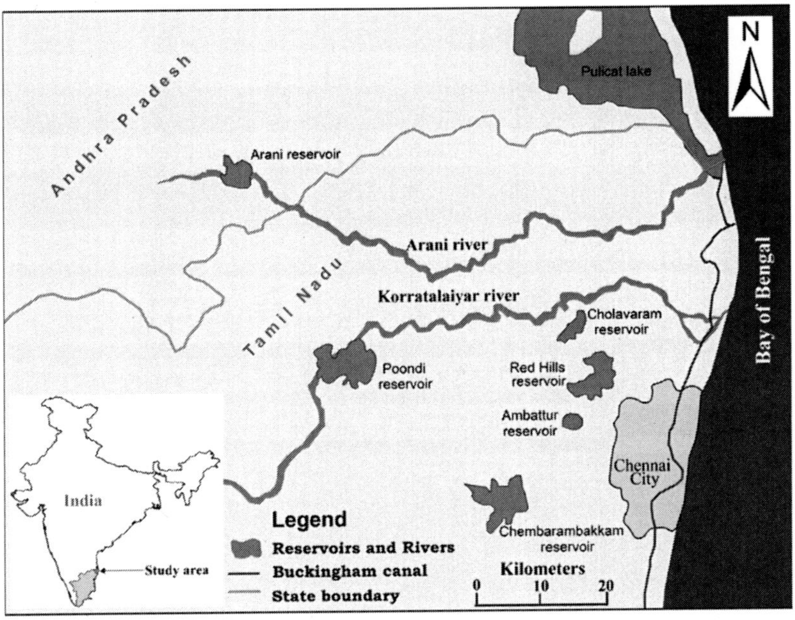

Figure 1. Reservoirs around Chennai (source: CMWSSB).

Conventional technical approaches to overcome the water shortages during summer were the construction of new reservoirs (e.g., Veeranam Lake Water Supply Project, commissioned in 2004), the increase of the capacity of existing reservoirs, and the provision of desalination plants. The Telugu Ganga Project diverts water from Krishna River in Andhra Pradesh to Chennai. Also, water pricing is practiced; however, in comparison with other cities such as Bangalore and Hyderabad only few households have functioning meters [15].

In addition, MAR has been practiced to replenish the aquifer and to mitigate seawater intrusion. Thereby, MAR was considered for replenishing the local aquifer at acceptable costs in order to "build a credit that can be drawn on in drought" [16]. Indeed, mitigation of seawater intrusion by MAR in the aquifers north of Chennai was observed by [17]. Thereby, in Tamil Nadu State RWH in all buildings is mandatory. Further, there has been a popular movement for the revival of traditional structures, e.g., Oorani for RWH or temple tanks for groundwater recharge. Two other technologies for MAR have been implemented: check-dams and infiltration ponds.

With respect to infiltration ponds there is one pilot study, implemented by Anna University. To be effective, a large number of small ponds would be required and a preliminary survey has shown that around 10,000 percolation ponds are feasible in the Arani and Koratallai river basin north of Chennai. Initial results indicate that approximately 40% of water stored in an infiltration pond may be recharged.

Similar figures about recharge were published for check dams [18]. Their overall storage capacity shall be 31 million m^3, with capacities ranging from 0.2 to 2.87 million m^3. Currently, there are nine dams at Arani River (4.42 million m^3), seven at Kortallai River (3.4 million m^3) and three at Palar River (5.18 million m^3). At Arani and Kortallai Rivers 71% of the planned capacity is implemented. Check dams at Palar River have lower priority, as the city depends only in extreme droughts on water from that river basin, which is at a distance of about 80 km.

Table 1 informs about the costs of recent projects.

Table 1. Capital costs for water supply infrastructure

Infrastructure	Water Supplied/ Recharged (Million m^3/year)	Capital Costs (Million INR)	Unit Costs (INR/m^3)
New surface storage reservoirs (Thervoikandigai-Kannankottai Reservoir)	28.31	3300	117
Increasing storage capacity of existing reservoirs (Cholavaram, Porur, Ayanambakkam & Nemam tanks)	16.08	1300	81
Desalination (Nemmeli plant)	36.50	8712	239
New check dam (Irulipattu check dam)	0.30	62	207
Infiltration pond (Anna University pilot project)	0.000175	0.015	86

Note: Data are drawn from project reports. These values were considered suitable for initiating discussion on stakeholder opinions, but should not be relied on for estimating actual costs of the respective types of infrastructure.

PROBLEM AND GOAL

This paper studies stakeholder perceptions in Chennai about groundwater management. It compares their views about three MAR options (roof top RWH in urban areas, large check dams and small infiltration ponds) with other approaches to overcome the problem of groundwater depletion due to over-exploitation, such as conventional infrastructure solutions (building or enlarging reservoirs), desalination, and non-structural policy instruments (e.g., water pricing). With respect to these options, the paper presents the interests, preferences and motivations of representatives of stakeholders at the national, state, municipal and individual levels, who took part in two workshops.

METHOD

A preliminary study identified the most relevant local stakeholders, in particular the governmental key-actors with interests in groundwater use, recharge, and quality (Table 2). Subsequently, the perceptions and preferences of stakeholder representatives for about six options to secure the future water supply for Chennai were explored, based on two workshops. Stakeholders interested in the topics (first workshop about options to secure water supply for Chennai, second workshop about infiltration ponds) were contacted and invited to participate. Thereby, "stakeholders" at the national, state or municipal levels are government agencies involved in water governance, while at the local and individual level these are groups of persons, companies or organizations in the Chennai area with a concern for groundwater related issues.

Table 2. List of stakeholders

Stakeholder/Institution	Level	Abbreviation
Government of India, Ministry of Water Resources, Planning Commission	National (Union State)	GoI
Central Pollution Control Board		CPCB
Central Groundwater Board		CGWB
Coastal Aquaculture Authority		CAA
National Green Tribunal		NGT
State Government of Tamil Nadu	Tamil Nadu State	GoTN
Public Works Department		TNPWD
Pollution Control Board		TNPCB
Water Supply & Drainage Board		TNWSDB or TWAD
Town & Country Planning Board		TNTCPB
Hindu Religious & Charitable Endowment Board.		TNHRCE
Water Resources Regulatory Authority (proposed)		TNWRRA
Chennai City Municipal Corporation	Municipality	CCMC
Chennai Metropolitan Development Authority		CMDA
Chennai Metropolitan Water Supply & Sewerage Board		CMWSSB

Food and mining industry	Local non-governmental	Industry
Private water companies		WaterBus
Tanker truck operators		Tanker
Water users associations		WUA
Agriculture sector		Farmers
Peri-urban villages		Peri
Peasants without own land		Workers
Residents of the city		Residents
Organizations of civil society		CSOs
Research centers and universities		Acad

A first workshop was conducted with participants from government and civil society. Also, several members of the project team took part and informed the participants about the present situation (see background information in Section 2).

In particular, respondents were informed about the technical options and costs observed for recent projects (Table 1): Increasing the capacity of existing reservoirs and groundwater recharge by infiltration ponds are most economical. The subsequent plenary discussions focused on the legal situation and on implementation aspects. At the end of the workshop, 25 respondents answered a questionnaire about the opinions concerning different groundwater management approaches, about the relevant criteria to assess these approaches, and about the opinions concerning different policy approaches.

The output of this workshop was used for the subsequent discussion of the legal and policy issues of implementing infiltration ponds in the area surrounding Chennai, mapped in Figure 1.

The project team presented these results at another workshop with representatives from government organizations and civil society. Again, at the end of the workshop 29 questionnaires were answered. (Respondents of the survey at the first workshop did not take part.) In addition to the previous questions about groundwater management approaches and criteria to assess them, a set of questions inquired specifically about infiltration ponds as well as legal and policy issues to implement them.

Participants of the workshops came from stakeholders groups, who could be decisive for MAR implementation. For the government (Table 2 for the abbreviations), these were members of the Chennai branch of CGWB for the central government; from Madras High Court for the jurisdiction, from several government departments (e.g., TNWSDB, Chennai, TNPWD, Chennai) of Tamil Nadu State; and from CMWSSB, Chennai, for the city. From civil society, there were representatives from business (e.g., consultants, advocates), NGOs (e.g., Alacrity Foundation, Chennai, DHAN Foundation, Chennai), and students and scientists from research institutions (e.g., Anna University, Chennai, Tamil Nadu Dr. Ambedkar Law University, Chennai). At both workshops, farmers took part, whose land might be used for MAR structures.

The surveys were conducted in the context of the future water supply needs of Chennai. For the interpretation, it should be recalled that the surveys were not intended to be representative opinion polls for any specific group, and *no concrete decision should be prepared*. Rather, these were explorative studies, where samples of 10–30 respondents suffice [19].

In particular, the sample was not representative for the population at large: 11% were women and the median age was 51, ranging from 20 to 37 for women and 25 to 71 for men. Further, 38% were from government or courts, 15% from research institutions, 19% experts from (other) NGOs, and 27% were farmers, some without education. To ensure their inclusion and to avoid misunderstandings of the questions, members of the project team (they did not take part in the survey) assisted the respondents in filling the questionnaires.

In order to identify explanatory structure, the survey data were processed by methods of pattern recognition, data mining, and social network analysis, using preferably non-parametric tests suitable for small sample sizes. The significance level was uniformly 95% (with the Bonferroni correction for significance of multiple comparisons, e.g., Milton Friedman's test). For one-sided 95% confidence intervals, Clopper-Pearson method (based on the inverse beta-distribution) was used, as it is conservative (higher confidence than stated as the nominal level). Software used was Microsoft Excel, XL-STAT of Addinsoft for statistical tests and data mining (an add-in to Microsoft Excel) and UCINET 6 of Analytic Technologies for social network analysis. In future decision making and project planning, accommodating and acknowledging stakeholder input and feedback will be important for a successful implementation and these methods may be applied again for such analysis.

RESULTS

Stakeholder Survey

This section lists 25 stakeholders (Table 2) and summarizes their interests in groundwater issues (Figure 2).

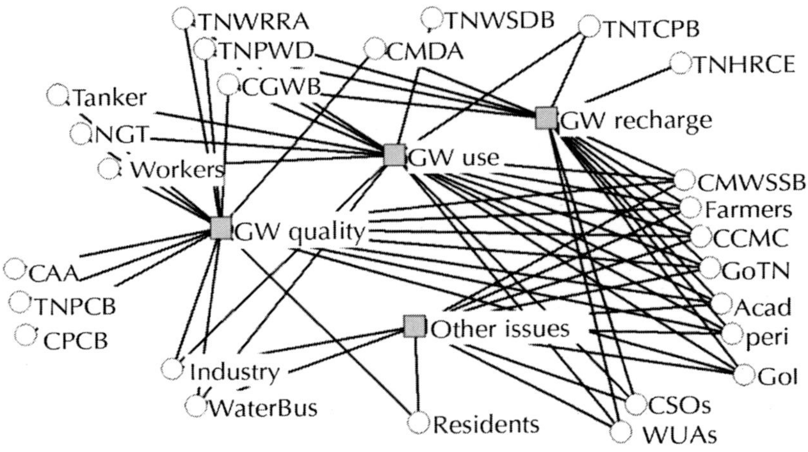

Figure 2. Stakeholders and their groundwater-related interests.

For governmental stakeholders, the interests are defined from the mandate (*i.e.*, laws and policies). For instance, as GoI delegated to the states groundwater responsibilities [20], and as the municipalities are responsible for actual water provision, at all levels of government there is an interest in all groundwater issues. More specialized government institutions have more restricted interests (Figure 2).

For institutional non-governmental local stakeholders (companies, universities, water user associations and other organizations), interests in groundwater issues, also indirect ones (e.g., industry, with direct interest in groundwater extraction, but not necessarily in recharge), were inferred from their business, research or other activities. For instance (with respect to CSOs), media regularly inform the public about the depletion of groundwater and sea water intrusion, and about advantages of groundwater recharge. The same is true for certain

NGOs, such as local groups against sale of groundwater in the villages around Chennai [21].

For groups of individual stakeholders (e.g., farmers, peasants, residents), the interests were figured out from their needs.

Comparing Acceptance for Water Supply Options

The results of this subsection are based on two stakeholder workshops and the subsequent surveys with 54 responses, of which 50 could be used, as all relevant questions were answered.

Stakeholders were asked about the acceptance of six approaches: increasing the capacity of reservoirs (representative of conventional approaches), desalination, non-structural policy instruments (such as water pricing), and MAR through RWH, check dams, and infiltration ponds. The options were chosen, because they were practiced or considered in the political discourse in the context of water saving and recovery, drinking water security, and groundwater recharge. (RWH is mandatory, reservoirs, desalination, and check dams are common, infiltration ponds and water pricing are discussed.) Further, they are typical instruments for different policy and technology approaches, and they operate at different scales.

Respondents were asked to assess the potential of the different options for securing the water supply (very high, high, low, very low) and to rank the options in terms of their individual preferences (from 1 = highest preference to 6 = lowest), using the "1224 competition ranking" (rank function of Microsoft Excel) to handle equals. From these answers, *low acceptance* (–1) of an option for a stakeholder was defined, if it was of low or very low potential and the ranking was five or six, and *high acceptance* (+1) was defined symmetrically (high or very high potential and rank one or two); the other answers were interpreted as *indifference* (0).

Summarizing the confidence intervals (Table 3), except for desalination plants and non-structural policies with low acceptance for at least ¼ of stakeholders, all other options appeared to be acceptable. Infiltration ponds were neither strongly supported, nor disliked by many.

Table 3. One-sided 95% Clopper-Pearson confidence intervals for acceptance of options

Option	High Acceptance		Low Acceptance	
	Lower	*Upper*	*Lower*	*Upper*
RWH	34%	59%	0%	6%
Enlarge Reservoirs	34%	59%	3%	17%
Check Dams	38%	62%	4%	20%
Infiltration Ponds	13%	34%	1%	12%
Desalination	14%	36%	34%	59%
Non-Structural Policies	2%	15%	28%	53%

Comparing also the distribution of the acceptance levels of each two options (Figure 3), desalination and policies had with 95% significance stochastically lower acceptance than RWH, building new check dams or enlarging reservoirs. For infiltration ponds (37 indifferent respondents) there were no significant differences in acceptance to any other option.

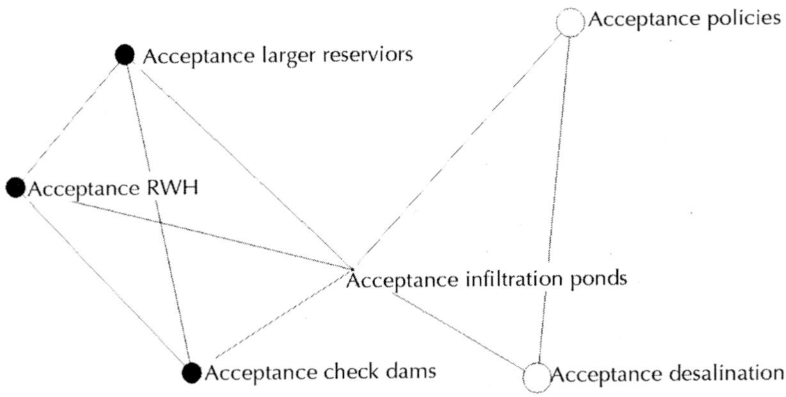

Figure 3. Pair-wise tests for differences in acceptance.

Stakeholder Motivations

The results of this subsection are based on two stakeholder workshops with 54 respondents.

Again, from 54 responses 50 were used, as they answered all relevant questions. To explore the motivation, respondents were asked to rank key-criteria by importance. As expected, on average human health (water quality) was most important, followed by the impact on the environment, social aspects (equity), impact on economy (costs, development), and practical issues (implementation, readiness of institutions). While between two consecutive criteria (e.g., health and environment) the difference was not significant, the criterion after the next one (e.g., social compared to health) had a stochastically higher (*i.e.*, worse) rank for importance (Friedman's test at 99.5% significance to correct for 10 pair-wise comparisons).

The 50 responses that answered all questions about the ranks of criteria and about the potentials, ranks, and acceptance of options were positively correlated, which indicates some consensus amongst respondents. A cluster analysis based on high correlations identified 22 respondents ("cluster respondents") with similar views (Figure 4). However, it also singled out 28 "non-cluster respondents", of them 22 idiosyncratic (no high correlation to any other response) and six almost idiosyncratic (highly correlated to only one other response). Regression trees (using XL-Stat) characterized the cluster-preferences (see [22] for atypical responses): Typically, 91% of the cluster respondents (20 of 22) had low acceptance for desalination and they ranked health first or second. Non-cluster respondents were expected to be more diverse, but typically, 68% of the non-cluster respondents (19 of 28) were indifferent or positive (high acceptance) about desalination and indifferent or negative (low acceptance) about check dams.

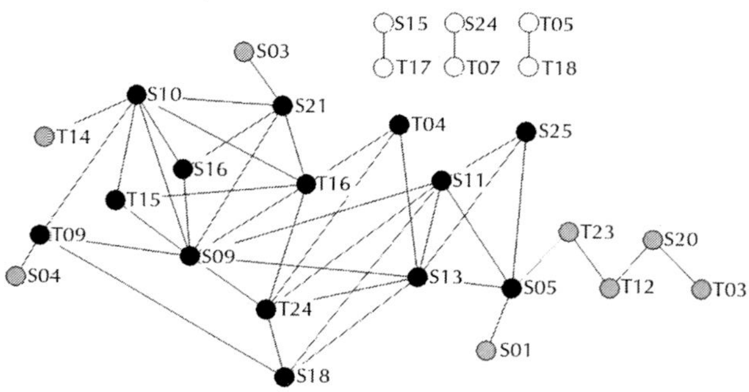

Figure 4. Cluster analysis of respondents to identify consensus.

Views on Legal Regulations and Policy Instruments for Implementing Infiltration Ponds

The results of this subsection are based on the second workshop.

For Table 4, 24 of 29 respondents answered all relevant questions (five did not). The workshop focused on the stakeholder views concerning the implementation of infiltration ponds, as in the view of the project team building thousands of small infiltration ponds would be an economically viable response to groundwater depletion, where the social group with the largest consumption of groundwater, the farmers, assume responsibility for its recharge.

Table 4 summarizes the views and displays significant differences between cluster and non-cluster respondents. The majority of respondents was critical about water supply, supported "the proposal to construct thousands of infiltration ponds in agricultural areas around Chennai" (interview question), whereby the farmers should take the initiative to implement them, the government should finance a substantial share of the construction costs, and farmers should operate and maintain their infiltration ponds and be responsible for the running costs (*i.e.*, for operation and maintenance). As to the differences between cluster and non-cluster respondents displayed in Table 4, cluster respondents have seen more responsibility in all aspects with the farmers.

For Table 5, the acceptance for instruments that support implementation of infiltration ponds was explored on the basis of 26 responses at the second workshop. (Three of 29 respondents did not answer all respective questions.) Thereby, the acceptance for policies was defined from the answers about the suitability (suitable, rather suitable, rather not suitable, not suitable) and the rank (1 = highest to 5 = lowest preference; respondents could propose as fifth category "other") of the policy instruments: High acceptance means suitable and rank one or two, low acceptance means not suitable and rank four or five. Table 5 displays the acceptance. ("Other" is not displayed, as only 7 of 26 respondents considered it.) Summarizing, two policy instruments to promote infiltration ponds were acceptable: supporting ponds using public funds and providing information. Thereby, "information" was discussed in a broader context of a (participative) communication strategy, as outlined e.g., by [23]. Making infiltration ponds mandatory for farms with more than one acre (about 4000 m^2) may be contested, with up to 32% opponents (not so much farmers) and up to 56% supporters. Rather not acceptable was fining farmers, who do not have infiltration ponds.

Table 4. One-sided 95% Clopper-Pearson confidence intervals for stakeholder views

Question	**Approval**		***Cluster Difference***
	Lower	**Upper**	
1. Water Supply: improvements needed	88%	100%	NC > C
2. Infiltration ponds: want them	71%	97%	No
3. Policies & laws support infiltration ponds	52%	85%	No
4. Farmers should drive pond development	32%	68%	NC < C
5. Government should drive pond development	21%	56%	No
6. Taxpayer should drive pond development	0%	12%	No
7. Consumers should drive pond development	3%	29%	No
8. Others should drive pond development	6%	34%	NC > C
9. Farmers should pay pond construction	11%	43%	No
10. Government should pay pond construction	40%	75%	NC > C
11. Taxpayer should pay pond construction	6%	34%	NC < C
12. Consumers should pay pond construction	0%	18%	NC < C
13. Others should pay pond construction	3%	29%	NC > C
14. Farmers should pay O&M of ponds	44%	79%	NC < C
15. Government should pay O&M of ponds	6%	34%	NC > C
16. Taxpayer should pay O&M of ponds	3%	29%	NC < C
17. Consumers should pay O&M of ponds	0%	18%	NC > C
18. Others should pay O&M of ponds	3%	29%	NC > C
19. Farmers should operate ponds	61%	91%	NC < C
20. Government should operate ponds	3%	29%	NC > C
21. NGOs should operate ponds	0%	18%	NC > C
22. Others should operate ponds	0%	18%	NC > C

Notes: Respondents could answer yes/no = ±1, and yes/no with reservations = ±0.5. "Approval" gives one sided 95% confidence intervals for the percent answering yes or yes with reservations. "Cluster difference" informs, if with 99.99% significance (Mann-Whitney test)

respondents of one cluster had a stochastically higher/lower approval and different mean approval rates.

Table 5. One-sided 95% Clopper-Pearson confidence intervals for acceptance of policies

Policy Instrument	High Acceptance		Low Acceptance	
	Lower	Upper	Lower	Upper
Public support for ponds	33%	67%	0%	17%
Information about ponds	26%	60%	0%	17%
Mandatory ponds (farms: 1 + acre)	23%	56%	5%	32%
Fine farmers without a pond	0%	11%	23%	56%

As the questions to identify needs for legal and policy changes were more specialized, respondents of the second workshop skipped certain questions depending on the expertise. (For this set of questions, 7% of 667 entries, *i.e.*, 29 responses to 23 questions, were not answered). The following percentages refer to those respondents that answered the respective questions.

- For the question, as to what institution should play an active role in groundwater conservation, most responsibility should rest with the State Government (48% stated that it would be the most important institution), municipal governments (for 48% it was the second most important institution) and farmers (for 33% the third most important group). Further, the State Government had (at 95% significance) stochastically higher priority than the Union Government (for 35% less important than the above actors or civil society/NGOs).
- For the question, as to whom governments should hear, when drafting and implementing water policies, farmers (for 68% the most important group to be heard) had with 95% significance a stochastically higher priority than all groups, except civil society (for 44% the second most important group). Surprisingly, courts (for 52% least important of all groups, except "other") had with 95% significance a stochastically lower priority than civil society and farmers.
- Most respondents (66%) considered the current situation as supportive for infiltration ponds and overall the present

groundwater recharge measures in Chennai would be adequate (50%). However, in view of the discussion at the workshop with respect to MAR, 59% answered that the current groundwater law was not adequate (see Section 5.5).

- With respect to groundwater-related institutions, 79% wished a law resembling the Tamil Nadu Groundwater Development and Management Act of 2003, which was never notified and finally repealed in 2013 (Groundwater Development and Management Repeal Ordinance). That Act would have foreseen an authority (TNWRRA) for MAR and Madras High Court repeatedly urged the state to notify it [24]. However, only certain aspects of this act were preserved in the form of Government Orders.
- Further, 69% would also approve of a law similar to the Model Bill for the Conservation, Protection and Regulation of Groundwater. This draft bill by the National Planning Commission of India is favorable for MAR.
- With respect to the characteristic features of these proposals, 78% support the establishment of a state authority responsible for water allocation. If there were such an authority, its agenda should include for 84% the regulation of groundwater extraction and for 79% the stipulation of MAR measures.
- Further, for 93% of respondents, a new groundwater law should be effective against encroachers who endanger groundwater. For 86% the legal regulations should be specific for regions with respect to MAR. 90% support stricter pollution control, where the local situation requires this. For 82% land utilization policies should be based on water availability.

Specific Observations from the Workshop Discussions

Compared to the other options, "non-structural policy instruments" was atypical, as it describes a bundle of policy instruments. In the workshops, the project team explained that this would include e.g., water pricing, banning or licensing of groundwater extraction, enforcing or supporting change to less water demanding crops, enforcing or supporting summer plowing to maintain soil humidity, or merely awareness rising amongst different target groups for issues related to water saving. However, perhaps as water pricing is practiced in Chennai (see Section 2), the discussions focused on "reducing demand by higher drinking water or irrigation water prices". Thus, for this paper "non-structural policy instruments" de facto means "water pricing and measures supporting it" (e.g., cut of energy subsidies, privatization).For the other options, no such problems occurred. Further, although respondents of the first workshop added several proposals for mitigating water scarcity, these proposals were conceptually similar to the considered options. Amongst the proposals was metering in apartment complexes and big hotels; control of demand by licensing; to encourage water saving toilets; recycling of grey water for domestic purposes (toilet flush); to simplify water recycling by separating wastewater according to its sources; clearing silt and sand from existing ponds to help sustain groundwater recharge; recharging storm-water and treated wastewater. Other suggestions were interlinking the rivers of Chennai and transporting water from distant sources.

For the criteria, additional questions (at the first workshop only) indicated that in applying the criteria to specific options, respondents lacked a common understanding about the meaning of the criteria. For instance, with respect to health, some approved of desalination, as it provides clean water, while others disapproved, as it does not provide natural water, which they perceived as healthy. Also for RWH, some were concerned about possible contamination, if collected rainwater was used for drinking, while others focused on other domestic uses and were not concerned. Similarly, for reservoirs and to a lesser extent for infiltration ponds, some were concerned about risks due to water contamination and dumping of waste.

In view of these experiences, the second workshop on infiltration ponds elaborated more on these criteria. However, with respect to the preferences there were no significant differences between the workshops, except for RWH: Participants of the second workshop had with 95% significance a stochastically higher acceptance for RWH than those of the first workshop (but at both workshops it was highly accepted). Perhaps, this was due to the focus of the second workshop on infiltration ponds, which are conceptually similar (small decentralized systems) to RWH.

For the legal situation, although by Table 3, RWH had highest support and least opposition, and at both workshops there was substantial criticism. Some stakeholder representatives disapproved of the mandatory implementation of RWH in every building without taking note of the different geological patterns, the different capacity of the ground to hold water, different rainfall patterns and complex groundwater usage. Stakeholder representatives of the second workshop therefore asked that regulations should allow considering the local situation (point 7 in Section 5.4).

These concerns about the consideration of the local situation apply also to the other options: If e.g., laws were requiring all farmers to build infiltration ponds, under certain circumstances such ponds may be meaningless.

Further, stakeholders reported implementation problems, as due to understaffing CMWSSB barely communicates with the public and lacks support from other stakeholders. This in turn results in deficient law enforcement: RWH structures are routinely monitored and maintained only in exceptional cases. Hence, stakeholders asked for more regular monitoring.

Similar implementation problems made current groundwater laws (point 3 in Section 5.4) inadequate: While CMWSSB denies groundwater extraction licenses for commercial purposes, the registration of wells largely failed and unauthorized extraction of groundwater is prevailing throughout the city; the offenders enjoy impunity.

DISCUSSION

The stakeholder surveys confirmed the known fact that a substantial fraction of stakeholder representatives was skeptical about desalination plants, which are amongst the most costly options to secure drinking water supply. Such low acceptance for desalination plants is known also from other countries, e.g., Australia [25]. In India cultural issues (also for educated populations, only spring water may be perceived as clean and healthy) aggravate this low acceptance problem.

The observed low acceptance of non-structural policies may be explained by the critical discussion of water pricing and privatization of water services. These policies are perceived critically also in other countries, e.g., Bolivia, where increases of tariffs have stirred violent public protests [26], Ghana and Tanzania [27], or South Africa [28].

There are concerns about environmental justice, as the burdens for the poor could be out of proportion [29].

Also, the high acceptance for RWH was as expected, as RWH is a traditional water supply option supported also by court judgments that repeatedly confirmed the eviction of encroachers from land used for RWH [30]. However, stakeholders had doubts about the efficient functioning of RWH structures.

The conventional approach to secure water supply is building new reservoirs. The stakeholder views on this option were not inquired, as there are limitations to new reservoirs, and to fulfill its water needs, Tamil Nadu state already operates reservoirs outside the state. This causes specific problems, as is illustrated by a recent interstate case at the Supreme Court of India [31]: Tamil Nadu state leases and operates Mullaperiyar dam in Kerala. Kerala was concerned about the earthquake-safety of the dam and enacted a state law to limit the reservoir level. In view of the consequently unmet water needs of Tamil Nadu, in 2014 the Court declared the Kerala state law as unconstitutional.

Increasing the capacity of existing reservoirs was the most economical of the considered solutions and it was generally accepted. However, stakeholders were aware that for reservoirs there is a need for regulations that consider the local situations: Vulnerable water bodies might need a higher protection than guaranteed by the national standards. A notorious example, for 15 years in courts, was the Orathupalayam dam project to use water from Noyyal River for irrigation, where five years after its completion in 1992, heavy water pollution from textile industry forced farmers to give up irrigation [32].

Groundwater recharge by check dams was the second most expensive option, but it was generally accepted and it is widely used. Conflicts about land acquisition plans may hinder the realization of such large scale infrastructure projects. This is exemplified by the delay of the construction of the Thirukandalam check dam [33]. For although landowners benefit substantially from check dams by increased yields [34], farmers fear receiving insufficient compensation for arable land that is used for such projects. Also, the survey confirmed that stakeholders were aware of the need to hear farmers, when formulating water policies (see point 2 in Section 5.4).

In terms of unit costs, infiltration ponds were second best with respect to unit costs. While the acceptance was not as clear as for the other options (see Figure 3), stakeholder representatives at the second workshop (about infiltration ponds) supported the idea to construct thousands of infiltration ponds in the rural areas surrounding Chennai (point 2 in Table 4). Farmers may at first not understand why they should give up arable land and spend money to build such ponds (just to secure the water supply of Chennai). Stakeholder representatives were aware of this problem and they approved the idea that the government should support the farmers in building infiltration ponds (Table 5 and point 10 in Table 4). Later on, the farmers should operate and maintain them without public support (points 14 and 19 in Table 4). Thereby, the implementation of infiltration ponds may also benefit from the observed high acceptance for RWH. Accordingly, infiltration ponds are small structures comparable to RWH structures and farmers will benefit from the aquifer recharge. Media reports [35] further emphasized that farmers may generate additional income from aqua-cultures (with risks for water quality).

Stakeholder representatives at the second workshop were not so critical about existing laws (many are used to apply them in administration and courts) and considered that current laws would support infiltration ponds (point 3 in Table 4).

However, in view of the workshop discussions, the majority, and also the project team, had critical views about the inadequacy of current groundwater laws and regulations (point 3 in Section 5.4). An example for the ineffectiveness of existing laws was the still applied national Easement Act of 1882 vesting owners of land with ownership of groundwater, irrespective of the rights of neighbors or public interests in groundwater preservation. Thereby, the interests of neighbors in water *de facto* have not been framed as legal entitlements or obligations. Further, national agencies (CGWB, CPCB) in charge of the implementation of national policies may not really influence actual decision making, as they tend to approve projects, which receive a "no objection certificate" from state agencies [36]. Yet, the stakeholder representatives considered that the national government should indeed have only a minor role for groundwater conservation, below state governments in importance (point 1 in Section 5.4).

For the specific problem of groundwater extraction, more than 75% of stakeholder representatives acknowledged the need to better regulate it and they supported the idea that a state authority should be in charge of MAR (points 4 and 6 in Section 5.4). Currently, different agencies of the government appear to act in an uncoordinated manner and without an integrated perspective about MAR [11]. For example, the water bodies and channels are not governed by CMWSSB, and neither are the temple tanks, which could serve as MAR structures. Another issue for the workshop was ineffective governance of groundwater, as commercial operators extract it unlawfully throughout the city.

The survey also identified a communication problem, illustrated by the lack of a common understanding of key criteria, such as health. A cluster analysis confirmed this lack of a common vision: While amongst 50 stakeholder representatives, 22 "cluster respondents" with similar preferences could be identified (Figure 3), the other 56% of respondents were almost idiosyncratic and perhaps unfavorable to MAR; e.g., the "typical non-cluster respondent" was indifferent or negative with respect to check dams.

CONCLUSIONS

Groundwater is an important source of domestic water supply in Chennai during the regular droughts and the peri-urban villages depend completely on groundwater. As agriculture and industry have been overexploiting groundwater, which is evident from the lowering of the water table and the intrusion of seawater, more effective instruments would be needed to control the extraction of groundwater and the use of water.

The paper investigated several feasible approaches, amongst them two MAR options, namely to build large check dams or many small infiltration ponds.

For the considered options, urban RWH is widely accepted and already mandatory, but stakeholders reported ineffective monitoring. Thus, better enforcement could make RWH more effective and better define the impact.

As to non-structural policy instruments, stakeholders identified them with water pricing and did not accept them.

Desalination plants and reverse osmosis of brackish water are too costly solutions to cover the basic demand, and consumers may not accept them.

Building new reservoirs for additional water or building check dams for groundwater recharge are costly, too, and in similar projects conflicts about land acquisition have caused substantial delays.

For the same reason, infiltration ponds could meet resistance, as thousands of ponds would be needed, but there is no legislation that would make them mandatory.

Further, for the implementation of infiltration ponds there is a coordination problem, as it would have not much effect, if only a few farmers would build small infiltration ponds: About 500 ponds would correspond to a small check dam and 10,000 to a large one. Thus, farmers would face costs, the groundwater table might barely rise, and if it rises, then farmers without infiltration ponds would be free-riders that benefit as well.

From these considerations it follows:

In the short term the most economical solution to secure water supply appears to be the enlargement of existing reservoirs. This solution is also generally accepted.

In the long term, infiltration ponds, which are the second most economical solution, are an alternative that most stakeholder representatives would accept. However, a coordination problem needs to be resolved.

All other options are already implemented (RWH; also most planned check dams along Arani and Kortallai Rivers are realized) or significantly more costly or not acceptable for most stakeholders.

To solve the coordination problem, stakeholder representatives support the idea to establish an authority in the state for licensing groundwater extraction and overseeing MAR. Accordingly, the establishment of a state authority responsible for groundwater governance and MAR (TNWRRA) would support the legal and policy measures needed to implement MAR structures. Thus, in this respect, stakeholders basically support the National Water Policies, where such instruments have been proposed. Of course, stakeholders did not envision merely another organization amongst the many existing ones, but wanted to see all groundwater responsibilities amalgamated.

ACKNOWLEDGMENTS

Co-funding to the collaborative project "Enhancement of natural water systems and treatment methods for safe and sustainable water supply in India—Saph Pani" (www.saphpani.eu) from the European Commission within the Seventh Framework Programme (grant agreement number 282911) is gratefully acknowledged.

REFERENCES AND NOTES

1. Starkl, M.; Brunner, N.; Zhong, Y.; Li, P.; Wang, Y.; Ericson, M. Overcoming barriers for the management of scarce water resources in Northern China. *Water Policy* 2014. in print.
2. Hardin, G. The Tragedy of the Commons. *Science* 1968, *162*, 1243–1248.
3. Ostrom, E. *Governing the Commons: The Evolution of Institutions for Collective Action*; University Press: Cambridge, UK, 1990.
4. Shah, T. Towards a Managed Aquifer Recharge strategy for Gujarat, India: An economist's dialogue with hydro-geologists. *J. Hydrol.* 2014, *518*, 94–107.
5. Sharma, S. *Managed Aquifer Recharge: Introduction*; UNESCO-IHE: Delft, The Netherlands, 2011.
6. Gale, I.; Dillon, P. *Strategies for Managed Aquifer Recharge in Semi-Arid Areas*; UNESCO-IHP: Paris, France, 2005.
7. Potter, E. Recycling Practice: Rainwater harvesting in Chennai, India, and the politics of water provision. In *Recycling Cities*; Inter-Asia Roundtable: Singapore, 2011; pp. 79–95.
8. Essl, L.; Starkl, M.; Kimothi, P.C.; Sandhu, C.; Grischek, T. River bank filtration and managed aquifer recharge as alternative water supply technologies for India: SWOT analysis. *Water Sci. Technol. Water Supply* 2014, *14*, 690–697.
9. Llamas, R.; Back, W.; Margat, W. Groundwater Use: Equilibrium between Social Benefits and Potential Environmental Costs. *Appl. Hydrol.* 1992, *1*, 3–14.

10. Jain, S.K. Harnessing the Managed Aquifer Recharge Potential for Sustainable Groundwater Management in India. *Water Energy Int.* 2012, *69*, 32–38.

11. Sakthivel, P.; Amirthalingam, S.; Starkl, M. A study on law relating to groundwater recharge in India. *Rostrum's Law Rev.* 2013, *1*, 83–101.

12. Douglas, I. Peri-urban ecosystems and societies: Transitional zones and contrasting values. In *The Peri-Urban Interface*; McGregor, D., Simon, D., Thompson, D., Eds.; Earthscan: London, UK, 2006; pp. 18–29.

13. Starkl, M.; Brunner, N.; López, E; Martínez-Ruiz, J.L. A planning-oriented sustainability assessment framework for peri-urban water management in developing countries. *Water Res.* 2013, *47*, 7175–7183.

14. Sakthivel, P.; Elango, L.; Amirthalingam, S.; Pratap, C.E.; Brunner, N.; Starkl, M.; Thirunavukkarasu, M. Managed Aquifer Recharge: The Widening Gap between Law and Policy in India. In Proceedings of the IWA World Water Congress, Lisbon, Portugal, 21–26 September 2014.

15. Brocklehurst, C.; Pandurangi, A.; Ramanathan, L. Tariff Subsidies in South Asia: Tariff Structures in Six South Asian Cities: Do They Target Subsidies Well? In *Water Tariffs and Subsidies in South Asia*; The World Bank: Washington, DC, USA, 2002.

16. Gao, L.; Connor, J.D.; Dillon, P. The economics of groundwater replenishment for reliable urban water supply. *Water* 2014, *6*, 1662–1670.

17. Parimala Renganayaki, S.; Elango, L. A Review on Managed Aquifer Recharge by Check Dams: A Case Study near Chennai, India. *Int. J. Res. Eng. Technol.* 2013, *2*, 416–423.

18. Nair, I.; Parimala Renganayaki, S.; Elango, L. Identification of seawater intrusion by Cl/Br ratio and mitigation through Managed Aquifer Recharge in aquifers north of Chennai, India. *J. Groundw. Res.* 2013, *2*, 155–162.

19. Isaac, S.; Michael, W.B. *Handbook in Research and Evaluation*; EDITS: San Diego, CA, USA, 1995.

20. Jamil, H.; Kumar, P.; Ismail, S.; Roy, R. Interstate Water Dispute and Federalism: Governance of Interstate River Water in India. *Civ. Environ. Res.* 2012, *2*, 11–16.

21. Butterworth, J.; Ducrot, R.; Faysse, N.; Janakarajan, S. *Peri-Urban Water Conflicts: Supporting Dialogue and Negotiation*; International Water and Sanitation Centre: Delft, The Netherlands, 2007.

22. There were 11 exceptions from the two major groups: Two cluster respondents (S01 and T24 in Figure 4) were indifferent about desalination; they had high acceptance for check dams, but (other than non-cluster respondents with similar preferences) only with "high potential". Three non-cluster respondents had low acceptance for desalination, but (other than cluster respondents) they ranked health at three to five. Six non-cluster respondents were indifferent about sanitation and they accepted check dams, but (differently from cluster respondents) with "very high potential".

23. Mefalopulos, P.; Kamlongera, C. *Participatory Communication Strategy Design: A Handbook*; FAO: Rome, Italy, 2004.

24. Madras High Court. *Pachai Perumal v District Collector of 08.07.2011*; Madras High Court: Chennai, India, 2011.

25. Dolnicar, S.; Schäfer, A.I. Desalinated *versus* recycled water: Public perceptions and profiles of the accepters. *J. Environ. Manag.* 2009, *90*, 888–900.

26. Bustamante, R. The water war: resistance against privatization of water in Cochabamba, Bolivia. *Revista de Gestión Del Agua en América Latina* 2004, *1*, 37–46.

27. Harris, L.M.; Roa-García, M.C. Recent waves of water governance: Constitutional reform and resistance to neo-liberalization in Latin America (1990–2012). *Geoforum* 2013, *50*, 20–30.

28. Flynn, S.; Chirwa, D.M. The Constitutional Implications of Commercializing Water in South Africa. In *The Age of Commodity: Water Privatization in Southern Africa*; McDonald, D., Ruiters, G., Eds.; Earthscan: London, UK, 2005; pp. 59–77.

29. Walker, G. Environmental justice, impact assessment and the politics of knowledge. *Environ. Impact Assess. Rev.* 2010, *30*, 312–318.

30. Madras High Court. *K.P. Vijayan v District Revenue Officer of 25.03.2008*; Madras High Court: Chennai, India, 2008.
31. Supreme Court of India. *Tamil Nadu v Kerala of 07.05.2014*; Supreme Court of India: Delhi, India, 2014.
32. Madras High Court. *M. Ramasamy v Noyyal River Ayacutdars Protection Association of 01.11.2012*; Madras High Court: Chennai, India, 2012.
33. Lakshimi, K. Long-Overdue Check Dam Project to Finally Begin. *The Hindu* 2013.
34. Sen, S.; Shah, A.; Kumar, A. Watershed Development Programs in Madhya Pradesh, Present Scenario and Issues for Convergence. In *Forum for Watershed Research and Policy Dialogue*; Institute of Development Research: Pune, India, 2007.
35. Renganathan, L. More Farm Ponds in Delta Will Shore up Water Resources. *The Hindu* 2014.
36. Koonan, S. Groundwater-Legal Aspects of Plachimada Dispute. In *Water Governance in Motion: Towards Socially and Environmentally Sustainable Water Laws*; Cullet, P., Gowlland-Gualtieri, A., Madhav, R., Usha, R., Eds.; Cambridge University Press India: New Delhi, India, 2010.

Citations

CHAPTER 1

Les Levidow, Daniele Zaccaria, Rodrigo Maia, Eduardo Vivas, Mladen Todorovic, Alessandra Scardigno, Improving water-efficient irrigation: Prospects and difficulties of innovative practices, doi:10.1016/j.agwat.2014.07.012.

CHAPTER 2

Ruth Meinzen-Dick, Property Rights and Sustainable Irrigation: A Developing Country Perspective, doi:10.1016/j.agwat.2014.03.017.

CHAPTER 3

Nicolas Greggio, Pauline Mollema, Marco Antonellini and Giovanni Gabbianelli (2012). Irrigation Management in Coastal Zones to Prevent Soil and Groundwater Salinization, Resource Management for Sustainable Agriculture, Dr. Vikas Abrol (Ed.), ISBN: 978-953-51-0808-5, InTech, DOI: 10.5772/50534.

CHAPTER 4

Mohammad Azizur Rahman, Bernd Rusteberg, Mohammad Salah Uddin, Muath Abu Saada, Ayman Rabi and Martin Sauter, Impact Assessment and Multicriteria Decision Analysis of Alternative Managed Aquifer Recharge Strategies Based on Treated Wastewater in Northern Gaza, doi:10.3390/w6123807

CHAPTER 5

Laurel J. Lacher, Dale S. Turner, Bruce Gungle, Brooke M. Bushman and Holly E. Richter, Application of Hydrologic Tools and Monitoring to Support Managed Aquifer Recharge Decision Making in the Upper San Pedro River, Arizona, USA, doi:10.3390/w6113495

CHAPTER 6

Robert G. Maliva, Economics of Managed Aquifer Recharge, doi:10.3390/w6051257.

CHAPTER 7

Basant Maheshwari, Maria Varua, John Ward, Roger Packham, Pennan Chinnasamy, Yogita Dashora, Seema Dave, Prahlad Soni, Peter Dillon, Ramesh Purohit, Hakimuddin, Tushaar Shah, Sachin Oza, Pradeep Singh, Sanmugam Prathapar, Ashish Patel, Yogesh Jadeja, Brijen Thaker, Rai Kookana, Harsharn Grewal, Kamal Yadav 4, Hemant

Mittal, Michael Chewand Pratap Rao, The Role of Transdisciplinary Approach and Community Participation in Village Scale Groundwater Management: Insights from Gujarat and Rajasthan, India, doi:10.3390/w6113386.

CHAPTER 8

Muhammad Arshad, Joseph H.A. Guillaume and Andrew Ross, Assessing the Feasibility of Managed Aquifer Recharge for Irrigation under Uncertainty, doi:10.3390/w6092748

CHAPTER 9

Norbert Brunner, Markus Starkl 1, Ponnusamy Sakthivel, Lakshmanan Elango, Subbaiah Amirthalingam, Chinniyampalayam E. Pratap, Munuswamy Thirunavukkarasu and Sundaram Parimalarenganayaki, Policy Preferences about Managed Aquifer Recharge for Securing Sustainable Water Supply to Chennai City, India, doi:10.3390/w6123739.

Index

R

S

T

V

W